Time: is the future already decided?
Another unsolved puzzle

Jonathan Kerr

Our two main theories are sending us opposite messages: relativity says the future is entirely pre-decided, but quantum theory has random events, that are undecided until they happen. Which is true?

Gb

Published by Gordon Books
9 The Green, Clophill, Bedfordshire MK45 4AD

The publication of this book establishes the author's origination of some of the ideas in it (and as intellectual property), as it precedes any other publication of them, except certain self-contained areas in peer reviewed journals by the author. The set of ideas in the later part of the book, in Part 19 and onwards, Planck scale time theory or rotation theory (known as PST or RT, also comprising DQM and PSG, from Books I and II), was discovered in 1995 and developed until 2023.

9th August 2024

ISBN
Paperback: 978-0-9564222-9-3
E-book: 978-0-9564222-7-9

With thanks to Nigel and Jenny Lesmoir-Gordon, their son Gabriel Lesmoir-Gordon, and to Johnny Wood and Paul Wheeler, for all kinds of help. And my thanks to Carlo Rovelli and Neil Turok, for very helpful conversations about dimensional quantum mechanics, Planck scale time theory, and areas of the background theory. Also thanks to the physicists I've had email discussions with, in particular to John Lucas and Ben Dribus.

Time is what clocks measure

Einstein's definition for time, 'what clocks measure', says very little, but it's by far the most used one in physics. It's well known that time is a mystery, but how we define it is a reminder - we don't know what time is.

Some links relating to Book I:

New theory of quantum mechanics shows matter is not in the eye of the observer, Sunday Telegraph: https://archive.ph/CxFol

Quantum mechanics' greatest puzzle 'solved' in a Surrey cottage
[Print version, Pressreader]

Paper: *An interactions-based interpretation for quantum mechanics*
https://doi.org/10.4006/0836-1398-33.1.1

Documentary *'The Interactions Avenue', partly a conversation between the author and Carlo Rovelli:* https://www.interactionsavenue.com

Time: is the future already decided?

Another unsolved puzzle

Introduction

There's a well known Einstein quote, often found on the internet, that reads: 'Reality is merely an illusion, albeit a persistent one'. This statement is very sweeping, and coming from Einstein, it can have a powerful effect. It seems to say that the entire world we find around us is unreal.

But it's a misleading misquote. His actual words were about time, not reality: they translate as *'The distinction between past, present and future is only an illusion, albeit a persistent one'*. The myth is quite widespread: allowing for a few different translations, Google finds the misquote in about 43,000 places online, and the accurate one in 29,000. There's even a place where the word 'albeit' has become 'Albert', but that's neither here nor there.

And on quotation sites one can find both quotes, each attributed to Einstein, but they're two versions of the same words. In a sideways sort of way, this backs up my point that they have different meanings. Somewhere along the line, the version with 'reality' must have appeared, split off from the original, and got into sources alongside it. But elsewhere, when not being misquoted, Einstein talks about reality differently - and perhaps not too surprisingly, as if it was real.

He's recorded as using words like that only once, a month before he died, in a letter sent to comfort the family of his lifelong friend Besso: *'In leaving this strange world, he has once again preceded me by just a little. That doesn't mean anything. For us convinced physicists, the distinction between past, present and future is only an illusion, however persistent.'*

Although it seems a bit of a tangent to look into the physics in these rather moving words, the idea of no distinction between past, present and future came out of one aspect of special relativity, which suggests motion through time doesn't exist.

Paradoxically, the equations describe the unexplained timeline motion - away from the past and towards the future - that we all seem to go through each day, along with all the matter around us. It's one of the main things special relativity does. In a way it's a theory *about* the timeline motion, as it showed for the first time how it varies its speed, allowing us to predict exactly how fast time will pass in all kinds of situations. But on close examination it also led, indirectly, to the idea that this motion can't be real.

Sixty five years after those words about past, present and future, the puzzle has come into sharper focus. We know from experiments that the timeline motion behaves as if it's real. Two objects can age at different rates, and get out of sync, leaving lasting age differences. This was expected, but explaining it without motion through time has turned out to be harder than expected. We also have major problems with time in other areas.

Nowadays physicists are in three groups on motion through time. One group think, as Einstein did, that it's some kind of illusion (or emergent), and after his lifetime a 'proof' was found that it's merely in our perception. But that's if spacetime is exactly as it was seen in 1966. Another group, growing rapidly since about 2005, and including some highly respected physicists, think this motion has a deeper basis that's not understood. But by far the largest group is physicists who say they're unsure about it. For one thing, *change* - which would also be an illusion - is so fundamental that removing it from physics is utterly impossible. There's no consensus on some very deep questions, and a lack of consensus often translates to 'we don't know'.

So time is still a central mystery. It was a mystery to the ancient Greeks, who laid the groundwork for our thinking. And although we have far better clues than they had, it's a mystery to us. The 20th century clues were weird, and accurate tests have confirmed them. But it's worth re-examining them in the light of 21st century physics, which alters how they look.

Recently the major drive in physics has been to fuse our two main theories, quantum theory and relativity. They disagree on time, so two different ideas of time must somehow be reconciled. This has brought the time problems to the surface, so a puzzle that for decades was set to one side has shifted from the edge to the centre, and become a key issue.

The problems are on the conceptual side rather than the mathematical side, and the solution may be too. Einstein and John Wheeler both said that in the future we'll find an underlying picture, that will explain what we know so far in physics. So from here a lateral jump may be needed, perhaps leading to some new idea or principle.

Lee Smolin is one of America's top physicists, and one of the most respected physicists in the world. His well known quote is: '*I believe there is something basic we are all missing, some wrong assumption we are all making. If this is so, we need to isolate the wrong assumption and replace it with a new idea. What could that wrong assumption be? My guess is that it involves two things: the foundations of quantum mechanics and the nature of time. [...] I strongly suspect that the key is time. More and more, I have the feeling that quantum theory and general relativity are both deeply wrong about the nature of time. It is not enough to combine them. There is a deeper problem, perhaps going back to the origin of physics.'*

And a few lines later: *'...time is represented as if it were another dimension of space. Motion is frozen, and a whole history of constant motion and change is presented to us as something static and unchanging. If I had to guess (and guessing is what I do for a living), this is the scene of the crime. We have to find a way to unfreeze time.'*

Attempts to unfreeze time - and its mathematical representation - have been getting increasingly adventurous recently, as people try to make adjustments to allow the joining of the two theories. In 2009 a theory that brings back Newton's view of time (at some energy levels) caused a lot of interest, and a meeting of top physicists at the Perimeter Institute. And in 2011, along with an international team, Smolin co-wrote a paper that tries to replace an idea that was hardly questioned for a century: spacetime.

Then in 2015, a groundbreaking series of experiments weakened the present view of time, in as far as there is one. It removed the single idea that led to several areas of it: the idea that the world is reversible at the particle scale. Most of our equations are reversible, which made some think that the world is as well at that scale. But time goes in one direction, so it seemed that this 'one way street' might be some superficial emergent effect, perhaps caused by thermodynamics, which also goes in one direction.

But we were guessing ahead of experimental technology, which then caught up. In 2015 it was shown that this imagined reversibility was false in the first place. Like the large-scale world, the small-scale world also turns out not to be reversible (Part 16). Something is creating an arrow of time at that scale, just as we find at all other scales. This has added to the increasing need for a new view of time.

Recently, what many non-physicists believe about time has been out of date. Some 20th century ideas about time, such as the idea that it 'might not exist', came out of one view of spacetime, but that view is rapidly disappearing. We used to see spacetime as fundamental, because it was about the dimensions. But this century, partly due to its incompatibility with quantum theory, many think spacetime emerges from something else, at present unknown. That's a major change - it's in Chapter 12. It leaves room for differences to the nature of time. But the public's view hasn't taken the implications onboard, and is a decade or so behind the physics community.

But it means adding an unknown layer underneath what we have, and that's enough to affect other unknowns. It means some of our earlier ideas about time, which we were never certain about, may be wrong. They could also be right, and they can't be ignored. But the goalposts have moved, so some very old questions might have new answers.

Not only do the public not know the answers, they often don't know the real questions, arising from the present picture. One question, and a key part of the puzzle, is the contradiction between our two main theories on whether the future exists. Quantum theory says it's unformed and unfixed, because of a deep-seated randomness about events at the particle scale. But special relativity suggests the entire future already exists, and is unalterable. Both of these theories are well tested - the mathematics of both has been confirmed by experiment many times. But it seems that our *understanding* of one of them must be wrong somewhere.

Nowadays there are more books that talk about time, but almost none that offer truly new solutions. Current approaches often attempt to find smallish tweaks *within* existing physics, to smooth the problems with time back into our present picture, in order to rescue it. So these books can be motivated to skate over the problems. But that tends to mean leaving out the best clues, because the problems are exactly where the best clues are to be found. And a minor fix may not be enough - if our view of time is wrong, an entirely new one might be needed.

Pulling all the clues together, including ones that are not mentioned, without assuming we must rescue a failing view of time, is what this book tries to do. It's for both lay readers and physicists. The aim is to show more of the puzzle than has ever been set out before: the real clues, the avenues of thought that lead off from where we are now, and what different *kinds* of solution might work.

The puzzle was a taboo until the late 1990s - some areas of it are so baffling that physicists were reluctant to talk about them even amongst themselves, let alone to the public. But in the 21st century there's a new openness, and we're questioning everything about time.

I hope the book sheds light on the mystery in one way or another - perhaps just by setting it out clearly, showing new aspects of it, and providing some food for thought.

But later it also gives possible solutions. The few avenues we have to choose from are looked at - all are shown to be blocked in one way or another. Then one of the blockages, via a lateral jump, is removed. The solution that opens up leads to a picture. It's a simple, unexpected new view of the universe, and a visually beautiful one, which fits neatly into many areas. It solves problems, and led to the solutions for all three puzzles in Books I, II and III.

Among the questions it then provides answers for is the surprising data from the James Webb Telescope, which shows galaxies forming far too early for us to explain. Some of our current theories - inflation and varying speed of light (VSL) theories - describe extremely rapid events in the early universe, solving problems like the horizon problem, but with no explanation. Events seem to have happened very fast in the early universe. (This is about real time rates, not cosmological time dilation, which is simply an observational effect, due to waves stretching as they travel through expanding space.)

The view set out here does what other views of time can't do at present: it shows exactly why 'fast early universe' theories work, and explains the JWST data. There's also an explanation for the mass discrepancy.

For anyone wanting evidence that there's undiscovered physics out there to be found, or a hint of the direction this book will take, Chapters 8 and 11 will give an idea, or page 237. And for physicists who understandably think the reasoning that led to block time can't be avoided, Chapter 43 has a scenario in which it doesn't apply - it's one of the most important points in the book. Page 91 has a list of nine clues, coming out of basic, well known physics, and they all point in the same unexpected direction. To me, taken together these clues make a bridge, that leads from what we already know, and on towards where we might go from here. And the very simplest part of the book starts on page 129, with a new, lateral picture of time.

JMK

The chapters in this book are short, averaging less than two pages. So the index is long, and is at the end of the book with the reference section. Because of this, the chapter index itself can be used to look things up.

Section I. The time puzzle

Part 1. An overview of the puzzle

1. The main paradox

There has been more confusion about time than in any other area of physics, even more than with quantum theory. It comes both from within standard physics, and also from those who try to go outside it. Before the 20th century, people struggled to understand time for thousands of years. Then we found a fascinating new set of clues, and later measurements confirmed them. But if anything, they leave us even more confused than before.

Nowadays we think time is an extra *direction*, a fourth dimension to go with the three basic space dimensions. But time moves on, and we don't know why - we all seem to move along the timeline each day. It's as if everything is moving in this extra direction, with all matter going through changes, and so moving along the time axis. The other dimensions don't have general motion along them, so it's unclear why this fourth dimension should have everything moving along it, or what this motion actually is.

But it may be that this motion doesn't exist. Fifty five years ago a 'proof' that it doesn't exist was found. It meant that if the spacetime geometry is exactly as we saw it in 1966, an event can be in the past for one person, but in the future for another. If so, the difference between past and future is apparent (in our perception), and not real. That meant there can be no 'now' moment, moving along between past and future - as there seems to be - and led to the view of the world known as block time, the 'frozen universe'.

With spacetime as it was back then, no-one could pull that bit of reasoning apart. It seemed that the motion is an illusion, or some emergent, superficial effect. But now we're not so sure. Nowadays we see spacetime differently, and far more people think time runs deep than fifty years ago.

So matter's motion through time is one of the real mysteries in physics. It's hard to see how it could exist, and it's hard to see how it could not exist. The illusion is thought to include all the processes of *change* we see around us, and the idea that the world is developing as it moves through time, with past behind it and future ahead of it. So the illusion, if there is one, includes a lot of what makes our world what it is.

Although these problems were set to one side for half a century, here in the early 21st century we need to solve them, and there has been a major focus on time in physics recently. At the very heart of the puzzle is this question of whether there's a present moment that moves along. The fact is, in our view of the world, a present moment that moves along through time is absolutely needed, and couldn't possibly be removed. Most of our physics depends on it - and yet on paper it doesn't exist.

And we find there's a lot on the table. Whether it's real or unreal, the things that depend on this timeline motion include the laws of physics themselves, and underneath them principles like cause and effect. Without some kind of flow of time underneath events, nothing could move through space - a mass couldn't have a speed in relation to another mass. And yet masses do move in relation to each other, and when they do they obey clear physical laws. So if motion through time is an illusion, a lot of what we call 'physics' would be the study of what this illusion does.

And if this motion is an illusion, it behaves in a surprisingly precise way. Its speed varies from place to place, slowing down and speeding up in different situations. And the speeds obey simple mathematical rules, which means we can compare them, and predict accurately how fast time will pass anywhere. We then confirm that by making measurements, and they come out right, to differences of billionths of a second. So this motion - that apparently doesn't exist - runs faster in some places than it does in others.

2. Fusing two theories

Since the 1980s efforts have been intense to bridge the gap between the two main theories, general relativity and quantum theory. The mismatch we find between them is a real problem in places. If we're going to unify them under a single umbrella, at least one of them must be changed. So there have been decades of efforts to build a bridge, and reach quantum gravity: theories like string theory are part of these attempts. But the surgery being attempted is mainly on the mathematical side, not on the side of ideas.

It may be that we're looking in the wrong place. There's reason to believe a deep conceptual mismatch underlies these problems, and we might need to solve that before making progress. The conceptual side of physics is about how we interpret things - it's our picture of what's going on underneath the mathematics. Some very good physicists have emphasised the importance of it. But looking at the conceptual side now, we're baffled.

Both the theories do well in experiments, but they have different pictures of time: time behaves differently in each. To build a bridge of ideas, two views of time must somehow be reconciled. That means we might need to solve a very old puzzle to move forward now, and it has brought it out, and to centre stage. It also means we probably need an idea.

And it could alter the landscape. That's possible now, as areas like spacetime, that were taken for granted, are open to question again. General relativity is entirely dependent on spacetime. And contrary to how it's often covered in the non-scientific media, general relativity is now widely seen as a large-scale approximation, and not valid at all scales, as some respected physicists have said recently. Raphael Bousso: *As a matter of principle, the theory is wrong. The seed is there. General relativity cannot be the final word; it can only be an approximation [...].* Sabine Hossenfelder: *Most physicists, myself included, think general relativity is only an approximation to a better theory.*

But special relativity, by contrast, isn't being questioned at all (in mainstream physics). The simple, more basic part of relativity, about motion in a straight line at a constant speed, absolutely has to be right. And yet the clash about time between the two main theories doesn't go away, it's still there.

This clash with special relativity is the interesting area. The puzzle is reduced to its simplest form, so we can get at it. And it's about the part of relativity that we're absolutely certain is right.

3. A disagreement about the future

Nowadays a growing number of physicists think some new idea or principle is needed, that will reconcile two different notions of time, and explain the contradictions. Lee Smolin, who I've quoted in the introduction, has set out what he thinks this missing piece of the puzzle should do. Although he says there's a deeper problem than the need to iron out the differences between two theories, he believes the way forward should 'unfreeze time', that is, our description of time. Taken literally, this would mean finding a way out of the proof of block time - the frozen universe - that came out of special relativity in the 1960s. But that has been very hard to do.

Some physicists have talked to philosophers over the last few years about the problems with time, having seen how deep the questions run that must now be dealt with, to move forward. But philosophers can't necessarily help if there's a key concept still missing.

The idea that the problem is somewhere in our underlying picture fits with the way the two theories disagree over whether the future exists. It's an odd situation. Quantum mechanics tells us of a fundamental randomness about events at the particle scale, and the large majority of physicists have believed that for eighty years. These random events mean the future is unformed and undecided, and doesn't yet exist. It leaves room for us to be shaping things as they happen, and so affecting the future, as we seem to be.

But special relativity, as we interpret it at present, seems to tell us that all of time is 'laid out' like a dimension - the future already exists in every detail, and is utterly unalterable.

So the two pictures we're left with disagree deeply, and in a way that's not negotiable - they simply can't both be true. And yet both theories have been tested for decades, and do exactly what they're meant to in laboratories. The mathematics of both has to be right. But it seems our *understanding* of one of them is wrong somewhere.

4. A state of flux

By the mid 1960s it was clear that special relativity, with certain assumptions that surround it, means there's no motion through time. It took sixty years to reach a rigorous proof, but the conclusion was well known before that. But paradoxically, although it can seem to imply no motion through time, special relativity is full of motion though time.

Special relativity describes a moving, shifting world, like the surface of some ocean. Time rates shift, the masses and energies of objects shift - things get distorted in relation to each other. And it always depends on motion. Objects are spread out across space in a network of shifting relationships, everything is in a state of flux, how things look depends on ones viewpoint and how one is moving - and everything keeps on moving through time.

And for some reason, *how one is moving affects how fast other things seem to be changing.* This network of shifting relationships might look strange and complicated, but the way any two objects change in relation to each other is mathematically simple, suggesting simple geometry, almost as if something straightforward was going on behind it all.

The entire world is like that. Even an everyday scene - an office with people moving around in it - has shifting time rates everywhere. These changes are often set out in examples with spaceships, as at high speeds the differences are larger. But they also happen at low speeds.

Whenever a hand moves through the air, its time rate is slowed slightly, and is altered by a factor near 1. If it's moving at 4 mph, the hand is slowed by a factor of 0.99999999999999982, which might not seem too important. But everything in the hand slows down, as seen by someone watching, whether or not the hand is attached to them.

The processes that are slowed down, and have their speed multiplied by that number, include the regular pulsing of atoms, all biological processes, even the ageing process. If we could sense these changes directly, we'd be aware of a level in our everyday world at which time rates constantly shift around. We can't, but in laboratories we watch them happening. We also know the simple mathematics they go by.

At first glance this looks like it *must* be an illusion. Two people passing in the office will each see the other slowed down in time. Although the two effects cancel, leaving no traces, each nevertheless sees the other in slow motion as they pass, and it's utterly impossible for each to be ageing more slowly than the other. So it seems there's an inconsistency between the two viewpoints, revealing some sort of optical illusion.

But now try making the situation asymmetrical: say one person walks off and returns. Someone who stayed put will have aged more, leaving a permanent age difference between their atoms. And a hand that moves through the air to reach for a cup will have aged less than the rest of the person's body. And again, age differences will remain.

So in symmetrical situations the time differences look like an illusion, but in asymmetrical ones they look real. So on the face of it, we have elements of illusion, and elements of reality all mixed up together - which shows what a strange conundrum it is.

To me that's the real test of any interpretation for special relativity. I've seen a few suggested pictures for what's going on, and have discussed them with theorists sometimes. The test of any interpretation is: does it behave like an illusion in symmetrical situations, but does it *also* behave like a real effect in asymmetrical situations. That's the underlying challenge: the need is for an interpretation that does both. And so far, none of them do - but I'll give you one later that does.

This puzzle has preoccupied us for a century, but after a while people found an apparent reason to think it might all be illusion, rather than this mixture of illusion and reality. And it seemed the illusion would include all the motion through time we seem to find around us. (This was later questioned by many physicists - the illusion idea started to be dropped in the '90s, and has been replaced by emergent time this century.)

Special relativity appeared in 1905, and by mid-century had been confirmed beyond doubt with experiments. Then the ideas surrounding it led to a major problem - it seems the universe is static and unchanging. People had argued that time doesn't exist decades earlier, but usually only by showing that we get nonsense out of our basic picture of time.

But in 1966 a rigorous proof of block time was published, coming out of the spacetime interpretation of special relativity. It implied that past and future were no different from each other, except in our perception. That seemed to mean there can be no moment called 'now' moving along between past and future, defining a difference between them.

The following year a respected American philosopher made it even stronger, and it then looked unavoidable. From then on, one of the greatest problems for physics was the fact that what we observe is very different from what we were getting on paper. That still applies now (and we still use the phrase 'on paper', but in those days it meant what it said).

In the world we know, the one that physics has the task of explaining, there's a timeline. People, things, stars and galaxies all seem to be zooming along that line. In fact there are many timelines, not just one, but anywhere you go in the universe there's a moment we call 'now' moving along. For an analogy, the 'now' is like the needle of a record player moving across a record, or the laser in a CD player. But what special relativity - as we understand it - shows us is only the record, with nothing to play it.

So at the moment we're not sure what we're looking at. Perhaps the timeline motion is unreal, and there's nothing playing the record except some mental illusion, or a superficial effect. Or instead, perhaps what we have so far is an incomplete, unfinished version of the world.

5. Do fish wonder about water?

Time was a mystery for thousands of years before the time rate variations were uncovered by Einstein. The Greeks wrestled with questions like how something can keep its identity, and yet be constantly changing as it moves through time. Because they found contradictions, they sometimes dismissed the idea of a flow of time.

In the 20th century the issues spread from philosophy further into physics than ever before, and by then we were dismissing the idea of a flow of time for entirely different reasons. The reasons with which the Greeks dismissed it probably wouldn't convince us nowadays - such as the essence of an object being unable to change, or an arrow being unable to fly because at any single moment it isn't moving. But even if the Greeks didn't arrive at the answers, they were the first to pinpoint some of the vital questions, as they pioneered their way into this and many other uncharted areas.

Time is so hard to define, and pin down, that it has very often seemed easy to dismiss it as non-existent. Sometimes less preferable, the alternative has been simply admitting we don't understand it. Things we don't understand sometimes end up under the carpet. But it's harder to dismiss it than it might seem - to argue that time doesn't exist, nowadays physicists are obliged to invoke illusions or undefined effects.

Some non-physicists don't see the need for this at all, or the difficulties with dismissing time. Instead they happily say things like 'time is a human mental construct'. The reason they say this is they've not been told the whole story. They've been told we think time doesn't exist, but not that according to our theories, *change* doesn't exist either: nothing moves or changes. That's more embarrassing - it's often not said. So many are left comfortably believing that time is a word we use, but that only change actually happens.

But if physicists could dismiss time that easily, they wouldn't take extreme measures, like invoking undefined illusions. The truth is, we have excellent mathematics for space and time, but we understand one far better than the other. And it's surprisingly difficult to take either road, and argue that time exists, or that it doesn't. Because problems are found in both directions, the fact that we haven't pinned time down is not evidence that it doesn't exist: for now it's simply something we don't understand.

Whatever the answer, time is hard to define because we live in it. To define something, you need reference points outside it. So like the famous fish who argue about whether water exists, it's hard for us to get a view on our world from beyond it.

But in reality, fish know about water - they sense the currents, and changes to its motion, with their fins. Birds know about the air, as they feel it shifting around with their wings. And since the 1970s we sense slight changes to the flow of time with our atomic clocks. This may not be the same, but it may be. It's a recent discovery, as the mystery is thousands of years old, and we don't understand it so far. We just have some strange new clues.

The clues from the 20th century pulled in two opposite directions. They really strengthened both views of time, leaving them in balance. On the one hand we found evidence that motion through time exists somehow - because the changes matter goes through in one place, happen at a different speed from those in another. Even processes with atoms, which have fixed unalterable rates, are slowed if the time rate slows, and can get out of sync. But we also found apparent evidence that the motion doesn't exist, with reason to think past and future are entirely interchangeable.

But to understand what physicists *actually think* about the timeline motion - people assume there must be two separate levels, that work alongside each other somehow. One has motion through time, the other doesn't. Many try to dismiss the problem that way, but later on I'll show that this doesn't work. One reason is that the two levels can't be two views on the same universe, as they disagree on something very important - whether the future is fixed and formed, or unfixed and unformed.

6. Bits of the world seen in slow motion

Time dilation is a rather beautiful effect, whatever it is. It's easier to measure than it might be, because the world is full of natural clocks. There are things at every scale that pulse with a regular rhythm, and the smaller ones are very accurate. Each runs at a characteristic speed, and slight shifts to the pulsing are happening everywhere. Whenever an object moves, everything onboard it slows down slightly.

Two objects that move differently can be brought back together, and they'll have new age differences. In standard theory nothing moves through time at all, so they can't have aged at different rates. Instead, the explanation goes: 'special relativity predicts it, and that's just the way it is'. It's often taken to be the odd behaviour of the time dimension. But that's questionable these days, as our view of spacetime has changed.

Along with time dilation due to motion, there's the kind due to gravity. If an object moves towards a mass like a planet, its time rate will be increasingly slowed down. Both effects have been confirmed by experiment many times. So a very large part of the mystery boils down to: *if an object is moving, or in a gravity field, it's seen in slow motion.*

The fact that there are two kinds of time dilation gives us a set of clues. We don't know how close the relationship is, but it's unlikely that there's no link at all. So there's more to go on than if it was a single effect. It's like parallax, a principle used in vision, and also in astronomy - two sightings on something give a far better fix on it than one.

So what do we know about the relationship between the two time dilation effects? Obviously anything that relates them might be important. One clue is never talked about - it may be partly because we don't understand it. But people never talk about the clues anyway. It's not as exciting as the clues in Part 2, which people never talk about either.

When an object's time rate slows down due to gravity, its energy decreases. But with the other kind of time dilation, due to motion, its energy increases. And interestingly, the decrease happens in exact proportion to the object's time rate, while the increase is in exact inverse proportion to it.

We don't know why either of these sets of changes happen, but this may be a glimpse of something. Both kinds of time dilation are exactly proportional to the energy, but going in opposite directions. If we knew more about what energy is, this might be easier to understand. (We know the various different kinds of energy are all part of the same thing somehow, but we don't know what links them.) Although I've never seen this point mentioned, it's a clue. A good solution for the time puzzle would probably explain both these energy relationships - later on I'll suggest one that does.

There are different quantities relating to both kinds of time dilation: velocity, mass, energy, frequency, time rate, distance from the centre of the gravity field, and so on. We know the exact mathematical links between them - all of them. We still don't know what's happening, but some people mistake these links for causes. For instance, some think the time rate variations in a gravity field are *caused* by energy differences.

But the other kind of time dilation goes the other way! The reality is, for now we can't assume more than that these effects happen alongside each other, and are linked. So far, we haven't been able to pick out cause and effect: we just have a list of parallel effects. But because we know the mathematics, it can seem as if we're recognising something. But we're not, because when we try to pin it down, deep contradictions appear.

7. What people think about time dilation

But not everyone thinks there's a cause to be uncovered. Standing back a bit, attitudes to special relativity - from which a lot of this arises - fall loosely into three groups.

Firstly, a small marginal group don't believe it, as it's counterintuitive. People in this group have often tended to question, deny or ignore the experimental results.

Among the physicists who take the experimental results into account, there's a large group who think that special relativity is just the way the world is. To them, our instinctive way of seeing things is false - the world is different from how our species intuitively assumes it to be. So they don't expect to discover underlying causes. They take a few basic assumptions on trust, and then find most of relativity stems from them. On the face of it this is a reasonable view - it certainly was in the mid 20th century, with the old view of spacetime. So perhaps we say 'that's just the way it is'. To me, this view makes sense, but only if one ignores some very important clues.

The third group is also large, and includes many good physicists. They think special relativity is right, but that the picture is incomplete, with more to be found out, especially about time, and perhaps some hidden explanation. The clues led me to this view in the early '90s (one in particular), and to be one of the people in the third group.

Loosely speaking, the three views above boil down to:

1. There's no puzzle because it isn't true.
2. There's no puzzle because the world just *is* like that.
3. There is a puzzle, and we haven't solved it yet.

The second view works better when applied to relativity and quantum theory separately, which is what physicists often do. Taken separately each can look like that, but take them together, and it's far less convincing. Standing back and seeing the wider contradictions between the two, such as the one about whether the future exists, is the kind of thing philosophers do, and physicists sometimes forget to do. But if one does stand back, the contradictions make it harder to say 'that's just the way the world is', because at least somewhere in our picture, it sems that that's just the way the world isn't.

View 3 can be mistaken for something else. This is an important point. It can look like a refusal to let go of intuition, but instead it's very often the result of sound reasoning. With special relativity, we have to let go of intuition, and all good physicists do - those who don't let go of it may take view 1 instead. But because one reaction to the time questions (which I've called view 3) can be mistaken for another ('clinging to intuition'), for decades good physicists who think there's a flow of time have been reluctant to say so, in case they look like that kind of intuition-clinging thinker. That's a very large part of why so many say they're not sure about the time questions.

From the overall picture, one basic point: physics is full of equivalence. Again and again we find similar mathematics describes more than one picture. We have some good mathematics, but we don't know what it's describing, and it's simple, enough for there to be a range of possibilities. That means what's underneath time might be something unexpected. It could be very different from what we imagine, or perhaps just slightly different.

The point is, *either* could remove our present view of time. Even a slight shift to the picture - a very minor alteration - and block time would go. The frozen universe would then unfreeze. In Chapter 43 I'll show a small but significant shift that does that.

But for now, I hope I've shown that we can't say we know what time dilation is. A lot of experiments have confirmed that it happens, but a good one was in 1977 at CERN in Switzerland. Muon particles were sent flying around the circular tunnel, using electromagnets.

The muons live around 2.2 microseconds before they decay and disappear, and this fixed lifetime was known accurately. But that's for when they're not moving - these muons were moving very fast, at around 0.9994c, near to the speed of light. So their time rate was slowed, and their lifetimes lasted 29.3 times longer than they would at rest.

This neat measurement confirmed the simple equation from special relativity that relates the speed of a moving object to its time rate. (The equation is at the start of Part 11: for this experiment, put in 0.9994174 for v/c, calculate t, the inverse of the result is 29.3.)

The numbers came out right in 1977 as well, and very accurately, as they do. We know what happens, it's just that we don't know why it happens.

Some think there's no answer, and the world just *is* like that. Others say that whether or not there's an answer, 'why' doesn't matter - only mathematics and predictions do. But that's a 20th century approach, and since then we've got stuck in some vital areas. To make progress this century, we might need to start finding out about the 'why' underneath these things.

Part 2. Hints at an unknown world

8. Something odd about energy

While on the subject of reasons to look beyond the view that goes 'that's just the way it is', let me outline a truly fascinating clue from another area. When you look closely at this, it shows you that something absolutely fundamental is going on, that we simply don't know about. So it's exciting to look at it. The clue comes, as clues do, from splicing two things together.

We know an object can have several kinds of energy. When it goes through changes, its energy gets converted from one kind to another. But afterwards the object's total energy, with the different parts combined, will be exactly the same. Changes like this happen easily, and are happening everywhere, all the time.

Richard Feynman pointed out how strange the conservation of energy is, in one of his famous lectures in the 1960s. He says it at least *looks* as if a deep unifying setup must link all the processes:

'There is a fact, or if you wish, a law, governing all natural phenomena that are known to date. There is no known exception to this law - it is exact so far as we know. The law is called the conservation of energy. It states that there is a certain quantity, which we call energy, that does not change in manifold changes which nature undergoes. That is a most abstract idea, because it is a mathematical principle; it says that there is a numerical quantity which does not change when something happens. It is not a description of a mechanism, or anything concrete; it is just a strange fact that we can calculate a number and when we finish watching nature go through her tricks and calculate the number again, it is the same. [...] It is important to realize that in physics today we have no knowledge of what energy is. It is an abstract thing in that it does not tell us the mechanism or the reason for the formulas.

He decides that there's no deep mechanism, and that it may be an abstract mathematical principle. But he finds it odd that it should be only that, saying that energy 'does not tell us the mechanism'. A lesser physicist might sweep this under the carpet, but Feynman exposes it as a mystery.

Nowadays it's still one, but in the sixty years since then, a lot has been found that unifies all matter, suggesting it has a common root, and is all the same at some scale. In 1963 Feynman didn't know about these later clues, such as from string theory, but they make an undiscovered mechanism underneath what he talks about far more likely.

Now we don't know what energy is (an article in the reference section shows how we don't). 'The ability to do work' doesn't say what it is. And Feynman says 'we have no knowledge of what energy is'. But just as a sum of money can be converted into five different currencies, but keep the same total value in money, energy can be converted into five different forms we know about, and still keep the same total value in *something*.

Some would say this total quantity is a mathematical thing only, as Feynman suggested. Well there are two possibilities - either it's that, or it represents something real. Since the total must be one or the other, it's worth searching for more clues, trying to see which it is.

And you find that this can be spliced with the mass-energy equivalence. The object's total energy often doesn't change, but if it does change, something revealing happens. The object's inertia also changes, and in exact proportion. Now inertia is a very chunky, physical, real world sort of quality. It's similar to an object's mass, and is, loosely speaking, about how much the object resists being bumped around.

We know that with any adjustments to the total energy (if the object gains or loses energy in any form), its inertia will follow suit exactly. The total energy somehow shapes, or perhaps is, the inertia. So it starts to look like the total refers to something real. It's hard to avoid, if adjusting it adjusts the inertia. So now we've hit a question. If the total energy is something real, *how do the different kinds of energy combine to make up the total?*

There are around fifteen different kinds of energy: chemical, electrical and so on, as well as the basic (unexplained) kind, that's contained in all matter. The changes from one kind to another happen easily, without affecting the total. The different kinds vary so widely that it's hard to see what could link them all up. But if the total is real, there must be something they all contain, just as gin, whisky, vodka, beer and wine all contain alcohol.

But it's hard to see what it could be - it must involve something fundamental, and common to all matter. What else do we have on it? Not much. It must be quantifiable, because we quantify it. And because it's involved in changes and dynamics, it might involve a mechanism or process of some sort.

9. What we observe

Bearing that in mind, now let's return to time dilation. I'm hoping that what

I've just pointed out will affect some thinking on time dilation.

When we look at the local variations to the speed at which time passes, we tend to surround our measurements with a landscape of ideas about what's happening. But where contradictions arise, and things are less than clear, it's good to strip things down to what we actually observe.

And what we observe is simple. When motion through time changes its rate in a particular place, something changes locally. Whatever it is, it affects all processes involving matter there. It happens across many processes, all going at different rates - the speed of each is changed by the same factor. It might be the regular pulsing of atoms in a rock, or the growth of cells in a plant. All the processes of matter change their speed, or appear to somehow.

When the speed of all processes is altered in a particular place, we say time has changed its rate. The word 'time' has no other meaning in that situation, it just means all processes. If time slows down, all processes slow down. The march of time means the march of all processes.

We have trouble defining time. But the slowing of what we call time affects *matter*, and it behaves like a real effect. It's very hard to see it as an illusion, because there are lasting results after the 'illusion' is no longer there. If two grains of dust move differently for a few seconds (it happens a lot), or spend time at different heights in a gravity field, we know that new age differences between them will appear that last billions of years, becoming a permanent feature of the universe, potentially visible from any viewpoint.

So whatever the effect is, it leaves traces behind it. It's not so much that this makes time dilation look real. It makes motion through time look real. If two objects turn out to have aged differently, it suggests that they actually aged, and have moved through time somehow.

This effect is happening constantly, across the universe. The 'twins paradox' makes it seem a rare exotic exception, which has helped people avoid what it tells us. Standard theory has no real explanation that fits, but bits of matter floating around get out of sync. It's as if an *overall speed* is adjusted locally. Do these local variations - to the speed of matter-related processes - suggest an overall mechanism involving matter, as that clue about energy seems to? It's worth thinking about. Could these two clues both hint at the same kind of thing, and rather reinforce each other?

Without that point about energy in the last chapter, time dilation might look like the dimensions shifting around, though we don't know why they should, and have had trouble assuming they do. But here we have another area that might suggest some general matter-related process. (It will look far, far more likely when the clues in Chapters 10 and 11 are added). What could it be? If such a thing exists, it might also explain why physics is so surprisingly linked up everywhere, with unexplained patterns and symmetries joining up all the different kinds of matter.

In the light of this, it's worth bearing something in mind. Every clue we have about what we call 'time' comes in the first place, directly or indirectly, from observing matter (and sometimes radiation). Matter can't be removed from the picture. It's impossible to study time without studying matter, because along with radiation, that's all we can ever observe anyway.

Because we only study time via matter, and because of the unknowns about time that have been there throughout, what we call 'time rate changes' are effectively about something that happens with matter.

But we don't think of them as being about the *nature* of matter. We usually think of matter as comparatively unimportant, and rather like a stick thrown into a stream to see what the current is doing. We assume that the current - the flow of time - is there anyway in some sense, though without a stick to throw in, we'd have no way of knowing.

But then we find we have to question this flow. And because contradictions keep arising, with motion through time indispensable in certain parts of our picture, but apparently not existing in other parts of it, we're in no position to assume very much. So when we measure the time rate changes, as we do, it's best to keep an open mind about what we're seeing.

10. The first two clues

In Part 2 so far, I've tried to show what kind of direction a solution could go in. Because we're stuck, and many think we need to solve the time puzzle to get to quantum gravity, it might have to be something radical.

I hope it has also shown the need for visual thinking: we've reached a point where in places, only pictures and concepts will help us to move forward. In quantum theory it's pretty obvious we need a picture, but having looked at the unknowns we have about energy in Chapter 8, perhaps it's clear there as well. From here, I'll outline three clues that point a finger at matter, from a list that comes later. The first is simple enough.

Basic special relativity says that when an object starts to move, three things suddenly change in a mathematically identical way: its energy, mass, and its time rate (relativistic mass is on page 38). Now *identical mathematics* implies a very close link. And one of them is time. So what could it be?

We don't know what the object's time rate is, but the other two are matter-related: its energy and its mass. They belong to the object. And if the object's speed varies, its time rate shifts up and down - exactly in step with them. So perhaps its time rate is also a matter-related property.

The next clue is that in quantum theory, which works extremely well, time is an unchanging backdrop, while in special relativity time shifts around. Both theories are well-tested: *they're both right*. So why two different pictures of time? There must be a way to reconcile them, with no contradiction. Perhaps both have the same unchanging backdrop of time, but one of them, special relativity, adds on top of that a matter-related mechanism, which shifts the time rate around locally. That removes the contradiction.

It might seem hard for a matter-related mechanism to do that. But the well-known 'light clock' used to demonstrate special relativity - hypothetical, but very buildable - is exactly that: a matter-related mechanism that causes time dilation, and with the right mathematics. So yes, it's possible.

These clues suggest time dilation is something to do with matter. Once you start to see the possibility, other reasons to think it a possibility appear. But these two points are not obvious before you see them. We're not 'geared' to search for a mechanism, as we already have an explanation for time dilation, which is deeply ingrained, and entirely different. It's actually very loose. It's that the effect is caused by the dimensions shifting together in a closely knit way, for some unknown reason, when an object moves.

But as in the chapter after next, in the 21st century we're moving away from the old view of spacetime, and starting to see spacetime as emergent, rather than fundamental - and emerging from something unknown. So what could this 'something unknown' be? What I'm saying here is it could be a matter-related mechanism. 'Something unknown' leaves a surprisingly large amount of room for new possibilities.

11. Confined light, the best clue

The third clue tells us what's really happening. It's the best clue in terms of the evidence it provides, but later on there's another one that's even better, in terms of giving a pointer in a particular direction.

Light and matter make the universe what it is. They're so important that we find out a lot just from looking at their similarities and differences. So if in a given situation one of them suddenly starts to behave *exactly* like the other, and all the differences disappear, it's going to tell us something.

Confined light often means 'light bouncing backwards and forwards in a box of mirrors'. When light starts bouncing, it suddenly takes on three properties of matter, which light normally doesn't have.

The first is mass. Light travelling freely has no mass. But it has energy: it's not surprising that in a box, light adds energy to the box. Everything has energy, the light and the box both had before we put them together. So it's not odd to find that one adds energy to the other.

But we know the light also adds mass to the box. That *is* surprising: free light has no mass, but confined light does. And the mass that it adds to the box is in the $E = mc^2$ relationship with the light's energy. So the extra mass belongs to the light, it's not something else.

Wait a minute. So bouncing light suddenly gets given *exactly* the amount of mass it would have if it was matter? Yes, it's well known, and agreed, though most physicists would avoid putting it that way. When light starts bouncing, its mass immediately goes from zero to E/c^2. This can be shown via standard physics - it's in the reference section, as are all points in this chapter.

Taking the form of standing waves has *turned the light into matter*, or made it mimic matter's behaviour very closely. It seems that the box is mimicking at a large scale what matter does at some extremely small scale, perhaps at the Planck scale, in some place where we don't yet know the physics. If there are what we call 'particles' at that scale, they may be made out of standing waves. If so, the box is behaving like a big, huge particle. So this starts to look like a major clue about mass, as well as matter.

Don't forget that mass isn't fully defined in standard physics. It's different in special relativity from in general relativity. And the Higgs field only deals with about 2% of the mass within an ordinary object - the elementary particles it contains. So mass is still very much unexplained.

But here we've found something that creates mass, and in exactly the right amount: *standing waves*. So it seems that matter's mass arises from that sort of setup - at some scale where we don't yet know the landscape.

Mass is the first of three properties of matter confined light suddenly takes on. The second is what we've just been talking about, also an effect observed via matter: time dilation. We think of time dilation as unrelated to matter, but when light starts bouncing, it takes on time dilation. This is as in the 'light clock' used to demonstrate special relativity, where light zig-zags backwards and forwards between two mirrors, and each bounce is one 'tick'. So our box has now become a light clock. If it starts to move, the clock will tick slower, as there's further to travel between ticks, on a diagonal path. This gives the exact mathematics of time dilation.

And if you place the light clock vertical on the ground, it has a third trick. This one isn't well known at all. The light zig-zags to the ground in the same time that any matter takes to fall the same distance. This is in Book II. It's one of two versions of a thought experiment in which light and matter released at right angles from the same height - light horizontally, matter to fall vertically - hit the ground at the same time, if they both do. Standard theory agrees to 8 decimal places, with no explanation. The rule was found in 2021.

So starting from a simple setup, just light bouncing in a box, we can generate three qualities observed via matter: mass, time dilation and a fundamental response to gravity. The first, mass, is well known - Viktor Toth, a respected physicist, has set it out. It's in a discussion *'Does light possess mass?'*, but the page is full of confusion and disagreement among physicists on how to take the issue. (Many don't see that the special relativity rules for matter, such as $E = mc^2$, are not necessarily universal, and assume we have to fit everything into a universal set of rules. They then have trouble doing that, and argue about it.) The second, the light clock, is also well known and is in Chapter 85. The third is entirely new, and in Book II.

So what does this actually tell us? It's not that confined light is like matter. What it tells us is that matter is like confined light. Among the ideas it begins to explain is time dilation, which being one of several ways standing waves mimic matter, is likely to be a matter-related mechanism. And that explains why the clues in the last chapter suggested the same. And *that* tells us what kind of paradigm shift we're on the lookout for, and what sort of direction physics might take for the next few centuries.

Why don't physicists see these points? They do - original thinkers do. Many take it for granted, but Gerard t'Hooft, the highly regarded Dutch physicist, co-wrote a paper that tackles questions usually avoided on light and matter. In the last three lines, he says: *One could say: "Matter is just 'canned' energy, a box with internal dynamics, and radiation is 'free' energy."*

Confined light is on the list of clues on page 91, but it was too important to leave out of the early part of the book. But now we go back to what we know about time, and the mystery, starting nearer the beginning.

Part 3. Existing ideas about time

12. Spacetime

Minkowski spacetime is our present interpretation for special relativity, and nowadays it's very much a part of it. It's also central to general relativity. It's the only main area of our overall picture that's untested, but assumed to be true. We don't normally do that, we like to test everything. Our view of time is *totally* dependent on it. For decades, weaknesses in that view have been suggesting spacetime is wrong. And recently its failure to fuse with quantum mechanics is suggesting the same thing.

The interpretation of a theory means our understanding of it, and our idea of what the mathematics is describing. Spacetime leaves a lot unexplained, but it pulls special relativity together, simplifies it, and works well in some ways. It's an incomplete interpretation, but it has got so intertwined with special relativity that many now think of it as part of the theory. And it can be taken in that way - as a reformulation of it.

A short time after Einstein's 1905 paper on special relativity, his old teacher Minkowski put together the spacetime interpretation. A few years earlier he had called Einstein a 'lazy dog', for skipping his lectures, and because of his lack of effort.

Spacetime is four-dimensional geometry. In it, time is an extra axis - a fourth dimension. Minkowski thought time is a different *kind* of dimension from the others. It was like the other three in some ways, different in other ways. He put all four dimensions together in an inseparable unified thing, then called a continuum, now called a manifold. Einstein didn't like the geometry at first, calling it 'superfluous learnedness'. He also said now the mathematicians had got their hands on it, he couldn't understand his own theory any more. But later on he came around to it.

Minkowski saw an analogy between space and time. The link was about how matter moves through them. There's already an analogy in human culture between motion through space and time - the first visual representation of time using space was at least 10,000 years ago. But special relativity seemed to elevate that link to more than just an analogy.

What strengthened the link was that one of these kinds of motion affects the other. Why this happens wasn't understood at all, but the result looked very geometrical, and Minkowski interpreted it using geometry. Officially making time a dimension led to a picture with interesting internal logic to it, but it wasn't clear why the dimensions should behave like that, shifting around in a closely-knit way when an object moves.

But underneath spacetime was the fact that these two kinds of *motion* seem very similar. One kind we know and understand well, the other kind we don't understand at all. And although we've never been able to pin the timeline motion down or even decide whether it exists, this similarity was a large part of why from 1908 onwards time looked like a real dimension.

In the light of that, it's surprising that the same picture, later on, led to the idea that this motion doesn't exist at all. The timeline motion was one of the foundation stones of the logic that later removed it. This doesn't mean that there has to be a flaw, but it does look suspicious - some philosophers have pointed out that concepts such as the idea of a present moment, which were used in the argument that removed motion through time, then suddenly didn't exist. But they tend to think the argument still holds nonetheless, and there's general agreement that spacetime leads to no motion through time. If there's a flaw, it's not located exactly there. Anyway, an analogy between two kinds of motion led to spacetime.

Spacetime is based on assumptions about time. We don't understand time, so they could be wrong. But it seems to join onto the theory in a reasonable way, and many think it must be true. It has got so entangled with the original theory that it's seen as part of it. People often assume it has been confirmed by experiment, but only the core of special relativity has.

Special relativity itself has been beautifully confirmed, and deserves all the credibility it gets. Spacetime often gets credibility that actually belongs to its neighbour. But although spacetime has led to real contradictions, it has been hard to imagine how things could be otherwise. Spacetime has also been extremely helpful in simplifying many other theories, and has entered into the mathematical landscape to such an extent that - until recently - there has been an enormous reluctance to question it.

But this century it's at last being questioned. I've quoted Nima Arkani-Hamed on spacetime before, but it's a quote from a highly respected physicist, and very revealing. The lecture was held at the Perimeter Institute. In his opening words, he said: *'Both these principles* [quantum mechanics and spacetime], *certainly one them, are going to have to be modified somehow, or changed in a very significant way, in our next description of reality. So I think almost all of us believe that spacetime doesn't really exist, that spacetime is doomed, and has to be replaced by some more primitive building blocks.'*

Later on I'll be doing just what he says is needed, having been working with primitive building blocks for some time (he can call them that, it's fine).

But what did he mean by 'spacetime is doomed'? Recently, incompatibilities between spacetime and quantum theory have got clearer. Because of those, also because of string theory, a growing number nowadays see spacetime as emergent, and from something unknown. But it used to be seen as utterly fundamental. Making something emergent, from an unknown deeper layer, is what we do with things that don't fit the picture too well.

But I hope that doesn't sound critical, as I think they're right about it. I think spacetime, or something similar, does indeed emerge out of something else. A lot of us take spacetime that way nowadays, but few know what they think it emerges from. But adding in something unknown leaves room for changes, to help us get past the problems with time. And they really should get solved while we're about it, as nowadays we're in the middle of giving spacetime a major overhaul.

After all, a lot of our ideas about time, which we take for granted, such as the idea that it can't be said to 'flow', came out of the earlier view of spacetime. But if time emerges from something unknown, that's not necessarily so any more. This is a real change: it needs facing up to. It means that spacetime, or whatever behaves like it, is going to need an explanation. We used to think it wouldn't even need one.

13. Block time: the frozen universe

> *Mary, Mary, quite contrary, how does your garden grow?*
> *Well time and change are rather strange*
> *Without some prior layer of change*
> *They won't emerge - I'm a physicist, but I can't say that I know!*
> *I just stand and watch amazed, to see it change and grow*
> *You might ask the gardener, but frankly, depend on neither....*
> *He hasn't a clue either!*

Minkowski spacetime leads inevitably to block time, and this was eventually proved beyond any doubt. If spacetime is *exactly* right, and as it was in 1966, block time is unavoidably right as well.

In the block time picture, the universe is frozen, and nothing moves through time at all. Instead, all of history is already laid out in a static sequence, like a long series of photos. The universe is a four-dimensional block, and just as a 3D cube has many 2D slices, this 4D block has many 3D moments. Each 'time slice' is like a snapshot of the universe. The slices all sit alongside each other, like those in the cube, spanning the history of the universe.

Spreading the block out a bit, it looks like a row of photos hanging on pegs from a washing line, each a little different from the next. There's no moment called 'now' moving through the block, and all these moments are equally important. This block is simply *there*, it has no development or change, and everything it contains, in every detail, is unalterable. Some birds in Hyde Park helped with visualising this the other day:

This strange, frozen picture is still officially the standard view. As I said, the block universe is like a record with no needle to play it. No-one has been able to go on anywhere from there, or genuinely explain why we seem to move through the block. We have nothing to make it behave like what we observe. Although it has given us enormous problems, and many have been unsure whether to believe it, for forty years it was hardly questioned at all. But since around 2005 a rapidly increasing number of physicists have been questioning it openly, including some very well known people.

If the block universe is like a cube with many slices, thinking of it that way, it actually looks less like a cube, and more like a flip book. That's because the slices in the cube are all the same, but the slices in the block are each slightly different from the next one. So what we have, for those who believe in block time, looks like a flip book, but with nothing to flip the pages. The idea that our minds are doing it fails to work in a lot of areas. So nowadays people are beginning to think there must be something physical we don't know about, that is doing the equivalent of flipping the pages.

This seems likely for a list of reasons, even though it would mean Minkowski spacetime is inevitably wrong. For one thing, the flip book pages seem very predisposed to being run in a sequence, because when they are, what we call 'sense' emerges. An orbiting object, if one tracks its position, will be a little further on each second. As its position changes, it turns out to be going by a physical law that can be written down. So not only does sense emerge when the frames are run in a sequence, consistent laws do as well. And because these 'snapshots' actually *are* being run in sequence somehow, in one way or another - it looks like more than an illusion is at work.

14. Illusion time

And yet we have what looks like a proof that nothing physical is flipping the pages, or moving us through the block - except possibly some process inside living creatures. What's known as the Rietdijk-Putnam argument seemed to prove beyond doubt that the difference between past and future is one of perception only.

But this 'proof' depends on some untestable assumptions, and an outdated view of spacetime. As I'll show, if it turns out to be wrong, it might contain an important clue, because the assumptions it depends on are few, and one of them would have to be false. These assumptions have seemed certain to be true for a century, but now we're not so sure. We can't check them, and we we always check things (that's what we're like!). But they've had to be taken on trust so far.

Although many have abandoned the illusion idea recently, it persists in some places. And block time remains the standard view, from a 'proof' that motion through time doesn't exist. This means that the motion is totally uncharted territory. When we measure local variations to the time rate, we're seeing an unknown phenomenon - it has no place in our current view of the universe. This isn't just because we think there may be an illusion, but we don't know what causes it. It's because whatever causes it, we have trouble fitting the illusion into our picture. In some places it really doesn't fit at all.

For one thing, because motion through time varies its speed from place to place, if we take it to be a perception-based illusion, we have to say that the illusion is getting 'slowed down' somehow. And it's then difficult to keep the number of illusions down to one. A second illusion is often brought in (like a four-dimensional perspective type effect) to distort the first one, slowing it down. We then find ourselves with two separate and interacting illusions, neither of which is understood.

But the trouble is, if an illusion that affects our perception is being slowed down, that shouldn't leave age differences between bits of matter behind it. Slow an illusion down, then speed it back up to its original speed again, and if it really was a perception-based illusion, everything should be just as it was before you started tampering with it.

But instead, altering the running speed of this 'illusion' leaves traces behind it. And this affects not only the ages of objects. It also affects what the block universe contains, if it exists. And that means what this frozen block will have contained all along, since way before (in our sequential way of seeing things) there were living creatures in it - who see illusions - at all.

And even worse, these creatures and the illusion they see were always going to appear, and go through the exact history that they do. And that goes right down to the smallest detail - everything was pre-ordained.

A close look at the block time picture, and it starts to look like the religious cosmology that predated scientific cosmology. It's very hard to make it into a scientific view of the world. But there have always been enough unknowns in the picture for people to feel that it might perhaps be that way.

These kind of questions are rarely discussed, and naturally enough, not many scientific papers are published on the idea that an illusion might be at work. People don't really *work through* the illusion idea - instead they decide, with some relief, that a psychological illusion is outside their field.

So like the old maps that had unexplored areas labelled 'here be dragons', our present picture of time has areas labelled 'here be illusions'. Some are prepared to fill in the gaps in the picture in that way, but if we do, there are still inconsistencies left over. Invoking unexplained illusions is an extreme measure for a scientist, but if these measures removed all of the problems, they'd look a lot better than they do. In fact, there's a list of deep-seated contradictions surrounding time in physics, and some of them remain even after dismissing large chunks of the picture as unreal.

For instance, the major disagreement between the two main theories about whether the future exists isn't removed by bringing in illusions. Trying to remove it in that way draws us towards dismissing much of what physics tells us as meaningless, and the cost of removing the problem just gets higher and higher. But it still threatens to remain, even after these desperate measures. So the illusion approach, even if we had a workable basis for it, doesn't even fix everything anyway.

And among the sacrifices we risk making are the laws of physics themselves, because many of the laws include some kind of flow of time, and they don't work without one. But some assume the time slices in the block universe are somehow set up with the right ordering to produce the laws, if they're run in sequence. Unfortunately, that makes it a universe that until living creatures arrive, only has physical laws that are implied - and then an illusion comes along and makes them work.

15. Emergent time

If the illusion idea was in many instances a 20th century broom to sweep the time issues under the carpet, the 21st century equivalent is the idea that time is an emergent phenomenon. By around the '90s, people started to see that the illusion idea doesn't work. So nowadays many take the flow of time to be emergent, which means something real but comparatively superficial, which emerges out of something more fundamental.

I've mentioned the growing number of physicists who take spacetime *itself* to be emergent. This chapter is about the more common, very standard idea of taking the apparent flow of time in that way. Like illusion time, it's a way to dismiss the problem, and like illusion time, it doesn't work.

Emergence is where some setup leads to what behaves not like the sum of its parts, but like a new whole, with new characteristics. Patterns created by the wind, such as sand dunes, are emergent - the resulting structure is more complex than the simpler forces that led there, and has a nature of its own. And the route there is so indirect that it's hard to predict what will emerge, just from the initial conditions.

With digital media, there's no shortage of examples of emergence. Photos emerge out of many pixels, music out of many samples. Musical notes taken one by one won't give an idea of the music. Looking at the pixels of a photo one by one, you can't guess the image. In this kind of emergence, the result is closely connected with what shaped the pixels in the first place. In physics it's different: what emerges is entirely new. Temperature is another example - it emerges out of the motion of many particles averaged over, and has no other existence.

When computers were invented, they allowed us to see emergence of a kind we never knew existed. Computers were trained to repeat a mathematical function again and again rapidly, and make the result into a colourful picture (the black and white one below will give some idea, or put 'Mandelbrot set' into Google images). People were amazed by the beautiful, visually engaging patterns that appear. These patterns are emergent - their existence depends only on a process of fusing things that arise at a separate level.

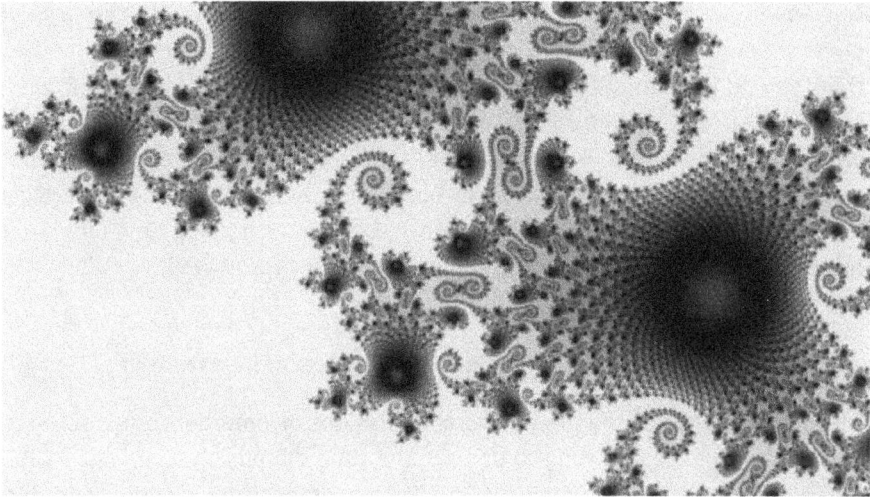

So could the flow of time be something that emerges? Taking the apparent flow of time to be emergent hits many of the problems of illusion time, but it explains far less without a psychological element. The original reason for the psychological aspect, which we'll get to, was that it seemed the same event can be in the past for one observer, but in the future for another. That made time look like something rooted in our perception.

But emergent time has trouble explaining that, and the idea also hits other problems. One is that it's hard to make something that moves and changes emerge out of something that doesn't. This question is the one in the poem a few pages back, about a garden growing. Some think time emerges from many small-scale events, but small-scale events need time already in place. That is, if anything actually happens, such as objects moving. If time emerges somehow, and creates a flow, or even just an apparent flow, it would have to involve a *process*.

But a process happens over time. The earlier examples of emergence involve a process: sand dunes from the wind, temperature from particle motion. Time is an ingredient of that. So if the emergence of time *itself* is a process, the trouble is, another flow of time would be needed underneath to allow this emergence to happen. As always, it's very hard for anything to happen if time doesn't get in there to make it possible. And we're so used to a flow of time existing, that many of our ideas about how it might emerge turn out to contain a flow of time already.

Now in some views of time, it's possible to include an underlying level. But emergence is definitely not one of them. If emergence needs a second flow of time underneath it, that defeats its whole purpose.

And again, just as with the illusion idea, there's a problem with the laws of physics. Some of them need a pre-existing flow of time to work (and the fact that they do work implies that one exists). But if that flow is emergent, basic laws such as laws of motion, which need time, would have to be emergent as well. Those who see time as emergent often expect to be able to keep other physics fundamental, and leave the problem behind. That's the underlying aim of that approach, and its real purpose. But in fact, making time emergent brings the need to make a lot of other things emergent as well.

16. The time dimension - just aspects of its nature?

Many people, rather than assuming that illusions or emergence are at work, think all of this is something to do with the time dimension. Ever since 1908 we've thought time is a dimension just like the other three in some ways, but different in other ways. The differences are major ones. As a result of this, there seems to be room for this dimension to be quite weird, perhaps weird enough to explain the contradictions. And on the face of it, it might be. But on closer examination, this idea starts to fall apart as well.

John Wheeler once summed up the relationship between time and the other dimensions in six words: *'Equal footing, yes; same nature, no'*. The 'equal footing, yes' is because time seems to be right there in the four-dimensional geometry along with the other dimensions. The 'same nature, no' is a real problem, because there's only one dimension of this kind, so there's nothing to compare time with.

And because of the differences from the three space dimensions, our *idea* of the time dimension isn't like anything else we know about. This conceptual isolation means there might be other differences we don't know about. And the differences we do know about could be incomplete - it's hard to tell if we have them all, and it's hard to tell if the ones we have are exactly correct. The time dimension is an 'ugly duckling' dimension, and not like the other dimensions in some ways. But in the mathematical picture these differences disappear, while in the conceptual picture they're large.

So during the early 20th century, when the list of differences from the other dimensions was growing, anything we didn't understand about time would tend to get put on the list. And *the actual nature of the dimension* became a bit of a dumping ground for the unexplained. It still absorbs whatever needs absorbing - contradictions, problems, questions. And despite all attempts to reduce the number of unknowns, we kept finding a lot of things that simply had to be accepted as part of this strange extra dimension. And it seemed we kept on having to say 'that's just the way it is'. But that leaves a view packed full of assumptions, and using a minimum of assumptions is one of our most basic principles.

The result was that during the 20th century, many set the time questions to one side, and put the contradictions down to something unassailable to do with the time dimension. Only now, because we need to solve the puzzle for another reason, has it come to the surface.

And looking at it now, because anything unexplained used to get added into the nature of the time dimension, we find that our picture of this dimension looks rather like something else. If all the qualities of the time dimension are written down, the resulting list actually looks less like a list of the properties of a dimension - it looks more like a list of clues.

Part 4. Matter - a stick thrown into a stream?

17. A mathematical link to matter

The next thing is to zoom in on some areas of the puzzle, and look at the role of matter in all of this. That includes a closer look at the first of the clues I've outlined.

Because you can only watch time at work by looking at matter (or radiation), matter's role could be more central than has been thought. During the 20th century matter turned out to have a far closer relationship with *space* than people had imagined. So perhaps it'll turn out to have an unexpectedly close relationship with time, in the 21st.

One interesting thing about the time rate changes is how local they are. Even an object as small as a particle has its own time rate, which can be worked out from how it's moving. Another particle nearby, moving differently, will have a different time rate. Both are made of matter, so we shouldn't rule out the idea that the cause is something to do with that. But if so, it would have to be something that happens at a very small scale.

According to basic special relativity, if a particle moves, its mass and energy increase - but not in proportion to speed. In the basic equations they change together, always c^2 apart, keeping the $E = mc^2$ relationship, as they do almost everywhere. Whatever that mass-energy link is, it would be risky to assume it comes apart.

But relativistic mass is counterintuitive, and some people want to remove it. Basic relativity has been strongly defended (Chapter 96), and is widely used as always. If an object is heated, it gains a little weight, as the particle motion causes a relativistic mass increase. One can calculate the mass it gains, but if you say mass *doesn't* change, mass and weight are no longer proportional, which is far worse. The point that follows still stands using energy alone, but I'll apply it here to both mass and energy.

When a particle moves, its mass and energy increase. They keep in step with its time rate. As the speed goes up, the mass and time rate also change, and stay in inverse proportion: each curve mirrors the other. So if the time rate is called t, the mass is $1/t$.

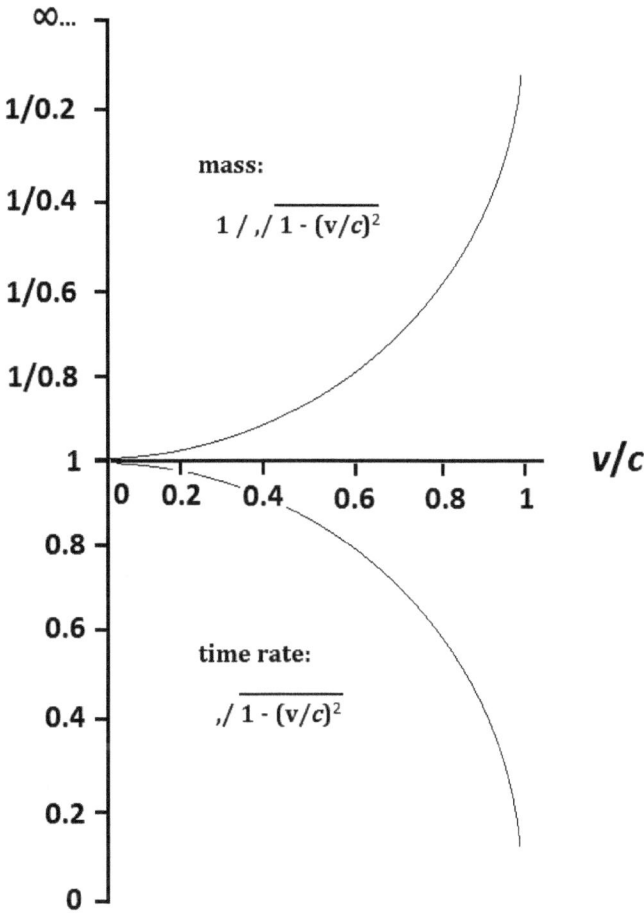

mass:

$$1 / \sqrt{1 - (v/c)^2}$$

time rate:

$$\sqrt{1 - (v/c)^2}$$

(vertical axis labels from top: ∞..., 1/0.2, 1/0.4, 1/0.6, 1/0.8, 1, 0.8, 0.6, 0.4, 0.2, 0)

(horizontal axis: 0, 0.2, 0.4, 0.6, 0.8, 1, labeled v/c)

For those who find the numbers of interest, take an object moving at 0.0447 of lightspeed. It will be seen with its time rate slowed down by 0.999, and its mass up by 1.001.

If it travels three times faster, its time rate goes down to 0.99, its mass up to 1.01. Three times faster again, its time rate goes down to 0.9, its mass up to 1.1. Both are getting further from 1, on each side of it.

On the horizontal axis those speeds, as fractions of lightspeed (v/c), start on the left, near zero. They're 0.0447c (13,400 km/sec), 0.141c (42,290 km/sec), and 0.4358c (130,670 km/sec).

And if the object gets up to 0.6c (about 180,000 km/sec) its time rate goes down to 0.8, its mass up to 1/(0.8), or 1.25 times when the object is still.

What's the point about this pattern? The time rate dances in step with these other changes, and the other changes are very much to do with matter.

So three quantities are linked. They shift together, but *all* the links between them are unexplained. Two have been shown in the lab to change in just that way, time and energy. With mass it's harder to test: one day we'll probably heat up some object and weigh it accurately enough. But mass and energy stick together via $E = mc^2$ almost everywhere else, and many physicists think basic special relativity is right, and all three do it.

If the object's speed varies, going up and down, all three wiggle together. It's not exactly with the speed, its with each other, as in those equations on the graph. It looks like whatever is being shifted around, it's some basic property of the particle, which closely connects time rate, mass and energy.

The idea that all three are linked in a way we don't understand is more likely, given that *we already know* two of them are closely linked in a general way - mass and energy, $E = mc^2$, without understanding the link at all.

This mathematical link time has with matter is even more interesting in the light of the problems with time. If we try to make matter *less* important, and like a stick thrown into a stream to see what the current is doing, and decide there's a time dimension that operates separately from the matter it affects, we run into contradictions. We find we have to question the existence of this 'current' that affects the stick, but is separate from it.

And then there's quantum mechanics. There too, matter seems to have one or two secrets still to be revealed, to put it mildly. There's also that strange point about an object's total energy, which suggests the same. And there are other clues - some are only hints, and can't necessarily be used to argue for anything. But rather like the butler in a detective novel, although often taken for granted, matter is beginning to look suspicious.

So if we started by making a list of ingredients needed to build a solution for the time puzzle, as well as the dimensions - which look necessary for quite a few reasons - we should put matter on the list.

18. Choosing the right question

So why is the world like this: why do we see bits of it in slow motion? People are often told 'That's just the way it is', by implication. But as Bruce Hornsby once said about that - '*Ah, but don't you believe it....*'

Even a ping pong ball, when it's moving, has its atoms pulsing slightly slower than the atoms in a stationary one. The question of why the ping pong ball's time rate slows down as it flies across the table is fascinating, but it may not be the right question to ask, to move towards an answer.

Another interesting aspect of time dilation is that if one treats the time rate as a literal speed in the mathematics (and it's far from clear why one should), *every object keeps a constant speed through four dimensions*. All matter then moves at the speed of light all the time, everywhere in the universe - if we look at four dimensions instead of three.

This means that matter seems to be constantly adjusting itself. If it moves a little along a space axis, it has to slow its motion on the time axis accordingly for some reason, to keep its overall speed constant.

This fact was discovered by Minkowski. It looks very general, but we have no idea why it's there. Although it wasn't understood, it played a large part in his decision to work with four dimensions, all inextricably bound up together, in the first place. In fact, our entire present view of time starts from this one unexplained point. In Chapter 62 I'll look into it, and ask what this constant overall speed might be. The chapter's called 'The most interesting clue of all', which may give some hint about my view on it.

Looking at things that way, the slowing of time now looks like a smaller part of some wider process, and an indirect result of it - noticeable if you zoom in on one aspect of it. So perhaps the time rate changes are adjustments that matter makes, for some reason, to keep its overall speed constant. It's well known that this way of seeing it involves the simple mathematics of speed components at right angles.

So instead of asking why the ping pong ball is observed in slow motion when it moves along one axis, perhaps we should be asking why when it moves on one axis, its speed on another - the time axis - is slowed in a way that neatly compensates for this motion. So perhaps the deeper question is: why does it keep adjusting itself to preserve its overall speed? And for that matter, why do all the other objects in the universe do the same thing?

Attempts to answer these questions are given a bit of edge when you remind yourself that officially, as far as we know so far, motion though time doesn't exist. That makes these changes all the more interesting.

Part 5. The wider arena

19. A less than obvious change

In the early 20th century science rapidly moved into a new phase. Measuring instruments got more accurate, and ideas more sophisticated. Among other things, we finally cracked into the world of matter at the particle scale. These changes led to so many differences from the past, that the most important one, perhaps, went comparatively unnoticed.

But there was a deep change in our view of interpretations. Before the 20th century, the hidden assumption was that explanations existed for everything we observed, if we could find them. This was part of the scientific approach that Newton and others had started, which brought with it, loosely implied, the optimistic and reasonable idea that every physical system was at least potentially understandable.

That view arose from the sudden breakthroughs in physical understanding that were being made everywhere, so it came from all directions. But Laplace in particular had crystallised it, with his picture of a universe that was like a repetitive clockwork-type machine. It moved according to simple, accurate laws, and in principle - with enough information - it was entirely predictable. This accessibility made it look understandable as well.

An interpretation is an explanation for what's going on underneath a bit of mathematics. It's often a picture - a visual picture made of concepts, usually linked to analogies. The principle thrives on similarities with other things we know about already, and it works because the world is full of things that are like other things. (In fact, mathematics uses the same kind of principle, and is also full of cross links and similarities.) When we have a full interpretation for a theory, we understand what the mathematics is describing.

Interpretations might seem like a minor side issue next to the mathematics - less rigorous and therefore less relevant. But that's false. When we make real progress, the picture drives the mathematics, and often shapes experimental physics as well. It's usually right at the cutting edge, helping us find our way. So the conceptual side has always gone hand in hand with the mathematical side, each shaping the other.

The 19th century was a time of enormous confidence, and seemed a golden age of physics. People felt science had arrived at a full understanding of the physical world, and what was left to do was mostly filling in the details. Until the last decade of the century, what was appearing seemed to make good sense. Light consisted of wave motion in a medium that filled all space, and even though not much had been found out about this transmitting medium, it seemed certain that it would be.

Maxwell's equations described electricity, and revealed that light was part of the same thing, electromagnetism. Although there was no interpretation for electricity, we were learning to use it, and had a mathematical description of how it behaved. Like a lot of things, it behaved very consistently, as if at root level something logical was going on underneath it, that would very soon be understood. And in the general scientific view of the world, it was implicitly assumed that everything had an interpretation.

But in the short time between 1900 and 1930, a devastating set of changes arrived, and these left people uncertain whether the physical universe could be understood at all. Relativity and quantum theory, which became our two main theories, both appeared. And both were so utterly weird in some ways that the question genuinely arose of whether we should even bother trying to interpret them. And both contained elements which, in entirely different ways, seemed to strike at the very heart of our view of reality.

But both could be used perfectly well without knowing what was happening underneath them, and to many it was far more exciting just to get on with building a new world. Quantum mechanics was a very accurate tool, which led to new technology. Attempts to Interpret the theory were put to one slde after a while, and left to the philosophers, who've been having a hard time with them ever since. And applied science got on with creating thousands of systems, including TVs and computers, that only worked at all because of the principles of quantum theory. They were used everywhere - we understood exactly how they worked at all but the very deepest levels.

In the early 20th century the Western world hit an identity crisis. It came out of a range of things, including the first world war, and the fact that religious beliefs were being eroded. But the arrival of these two vital and apparently uninterpretable theories was also a real factor, contributing to a feeling that there was less to believe in, and less meaning and sense behind the universe. Theories are also about the universe's knowability, so there now seemed less meaning and sense behind our relationship with it as well.

One of the important concepts behind quantum mechanics was Heisenberg's uncertainty principle, which speaks for itself about how it made people feel. Contrary to the neat, ordered picture imagined by Laplace, the universe had turned out to be random and unpredictable in some fundamental way. And it seemed to be one with which we couldn't negotiate - science had reached a boundary, and from there it was all just probabilities.

At the time there seemed less reason to believe in humanity, but also less reason to believe in anything outside humanity - which doesn't really leave very much. Then in 1931 Gödel showed that even our mathematics has to be questioned, and is a skin that doesn't quite cover all of what it describes. The cracks were showing in many of the apparently reliable structures around us, and the feeling was that we were now losing our innocent illusions.

So we let go of the underlying assumption, often taken for granted before, that everything we observe has an interpretation, just waiting to be found. It wasn't that people decided this idea was wrong, but it had become clear that it wasn't necessarily right, and that was a major change.

But interpretations are a vital part of science - this can be shown easily. The same theory can have two interpretations, both of which fit the mathematics equally well. But they can turn out to have different consequences, and lead to different experiments, or different results for one experiment. It's not that there was much choice about it, but carrying on without interpretations is like trying to do a jigsaw blindfolded. One can still test out ways of fitting the pieces together, which is like the mathematics. But there's an urgent need to see the picture as well.

20. Find the picture

To some physicists nowadays the conceptual side is not too important. Some think only the mathematics of a theory matters, along with its confirmability by experiment. And nowadays theories are sometimes arrived at by guessing at the equations, then looking at computer models to see if it works. This can help to narrow things down, but it's a young method, and we don't yet know if it can lead to major breakthroughs.

And we've tried to forge ahead without a picture. The mathematics we have fits well with the real world in some areas, but in others we lack a link to the real world. What we have is incomplete at best, we don't really know what's underneath our theories. Some say interpretations don't exist for them, but in the past whatever we've struggled to understand has usually turned out to have an explanation, with a century or two of progress.

Either way, few would disagree that if interpretations exist, we need to find them. But that's a lot more urgent than it seems - in the present situation we might need the picture to move forward. Not everyone thinks there's a need for conceptual thinking to make the next step, but that view has become *far* more common than it was twenty five years ago.

It's a need that may also be at the root of problems in cosmology. We might need a better understanding of our own theories before they can be applied to the universe at large. We've 'boldly gone', with excellent new technology to see what we can see out there, assuming that our present theories will be enough.

But we keep finding major discrepancies. So far we're still plugging the gaps with confidence - dark matter, dark energy, anomalies in the early universe, now confirmed by the James Webb Telescope data. But the truth is, we're in a very new game, and no-one knows how things will settle in. We might be twenty years from a picture that truly works, or several hundred.

So there's a need for conceptual progress. And although the puzzles with the main theories were not solved using the clues that accompanied them at the time, new clues have arrived since then. In the late 20th century Planck scale physics brought new possibilities - if you like, new pieces of the jigsaw are in front of us, and they might make all the difference.

In the past we've made progress using provisional conceptual pictures, which are later replaced by better ones. This has been possible because physics is full of equivalence - the same mathematics can describe more than one thing or process. This *interchangeability between concepts* has allowed us to move forward in a series of steps, replacing one picture with another as we go. And that in itself suggests, given the inconsistencies in our view of time, that the current landscape will have to change. But the interesting thing is, this leaves room for John Wheeler's 'simple idea' to exist - the final picture, perhaps to be found when the last step is made.

So the way forward demands new interpretations. But unfortunately, people block that path. Special relativity works well, but its interpretation doesn't. And because the interpretation has problems, it has been pushed to one side - and interpretations have been pushed to one side.

People often explain one bit of special relativity by referring to another part of the theory, giving an explanation from within that same set of ideas. The implication is often 'that's just the way it is'. But the mathematics is simple, and of the kind that might arise from a range of physical landscapes. So there could be a landscape out there (literally 'out there') that fits the clues more closely than anything we have at present.

There used to be a reluctance to discuss the time issues, but recently they've been faced up to a lot more. But people have been slow to acknowledge that what we have in front of us is a *conceptual puzzle*. My own view, since the '90s, has been that we need to turn round and look at the conceptual side. We're now so stuck that there's a new openness to that idea.

I found that view expressed by Oxford University physicist David Deutsch, in a New Scientist magazine article called 'Crunch time for physics', which was about new ways forward. The article called Deutsch's view 'radical', which is that we must abandon an approach that has worked for a century: the idea that one can start with mathematics, and expect to get to reality that way. The right order in the present situation, according to Deutsch, is to look first at the problems in our *understanding* of the world, then think what changes to our view might lead to solutions, then get the mathematics, then test the theory. He says we often try to get to the mathematics first, but in his view, in the present situation, that simply won't work.

This book came out of believing that for twenty five years, and is about what was found by deliberately looking away from the mathematics - but referring to it very often nonetheless - and examining the puzzles from the conceptual side. And the conceptual view doesn't only paint pretty pictures: it leads to new mathematics.

Nowadays physicists are searching through existing mathematics, looking for a shortcut to the step after next. But you can't. In this situation, you can't get there without an intermediate step - we need the picture. Searching through the mathematics can be like looking for a needle in a haystack. But (as can be seen in Appendix D, '*The proof of the pudding*'), having the right picture can be like parking a van with a large electromagnet next to the haystack - what you're looking for flies out.

21. Towards a theory of everything

There has been discussion about what a theory of everything might contain. John Wheeler said the final breakthrough would not be a mathematical one. He said: '*To my mind there must be, at the bottom of it all, not an equation, but an utterly simple idea. And to me that idea, when we finally discover it, will be so compelling, so inevitable, that we will say to one another, "Oh, how beautiful ! How could it have been otherwise?"* '.

There are so many different versions of this quote, one wonders how many it started from. But an archivist I've talked to says Wheeler made the point 'at least a dozen times' in his lectures and notes, in various words. Einstein said he believed '...*the principle of the universe will be beautiful and simple.*' I've mentioned what these two physicists said a lot. They were visionaries, telling us what to look out for, and we should listen to them.

Einstein expressed the need for a conceptual basis in the last sentence of his last article: '*The last, quick remarks must only demonstrate how far in my opinion we still are from possessing a conceptual basis of physics, on which we can somehow rely.*'

Six years earlier, in 1949, he said that such a basis will be found in the future. And he gave a hint on where to look for it, as he was talking about a difficult question, how to specify matter's position in quantum theory: '*It may appear as if all such considerations were just superfluous learned hair splitting, which have nothing to do with physics proper. However, it depends precisely upon such considerations in which direction one believes one must look for the future conceptual basis of physics.*'

Some think we're one step away from a theory of everything (TOE), and that the final step is a mathematical one. But one look at the contradictions with time shows that an idea is needed - no mathematical breakthrough would be enough to fix them, as they're in the conceptual department. But it's exciting that at any time these days, we might find Wheeler's 'utterly simple idea'. It has been so since the 1980s, when we reached the Planck scale, which made finding a TOE much more of a possibility.

What would this theory look like? To many it's about unifying the four forces. Progress in physics brings things together under fewer headings. Electricity and magnetism were unified in the 19th century, and electromagnetism, the result, was unified with the weak force in the 20th. Three of the four forces have ended up in the standard model: electromagnetism, the strong nuclear force, the weak force. In each case, the particle that 'carries' the force has been predicted and detected. That whole approach has been very successful, and many think it's heading towards a big TOE.

But the standard model is based on quantum theory, and the mathematics of that is complete already. It was shown to be in a paper from 2011, which set out a near-proof that any extension of the theory can't make the predictions more accurate. So we have the mathematics.

It's interesting to ask what a theory of everything would have to say about that. A few paragraphs back, Einstein was seen looking at a similar question, in 1949. If a TOE was only mathematical, it would have nothing to add to our present knowledge of quantum theory. That area might already be in place, though it might somehow link to there mathematically from other areas. But many of us think the final theory should do more than that.

Surely it should explain what's going on underneath the observed behaviour in quantum mechanics - and in particular the sudden change from undefined waves to well-defined particles. That change must have something going on underneath it, and the area known as decoherence has now started to show us that it has, from both theory and experiment.

We're only guessing on this kind of question, of course. But if a TOE would actually *explain* quantum mechanics, as many of us feel it should, it's going to be conceptual as well as mathematical - just as Einstein and Wheeler said it would be.

22. Hidden simplicity

Interpreting physics is like trying to explain the few bits that can be seen of a partly hidden structure, like a submerged object that breaks the surface of the water in places. Wherever it's visible above the water, there will be clues about what the whole structure is. Once we know what the submerged parts look like, the visible parts should make far better sense.

If a backdrop picture could be found that does that, it would have to explain what's currently visible in physics. Is it possible that some hidden structure connects them all? A lot of physicists think it, but only people like Einstein and Wheeler feel in a position to say it.

But some don't expect a simple, clear solution. David Gross, whose work has been at the forefront of string theory, and won a Nobel prize in 2004 for his work on the strong nuclear force, said in his closing address to the Solvay conference the following year: *'The state of physics today is like it was when we were mystified by radioactivity. They were missing something absolutely fundamental. We are missing perhaps something as profound as they were back then.'*

Elsewhere he says: *'The real change that's around the corner* [is] *in the way we think about space and time. We haven't come to grips with what Einstein taught us. But that's coming. And that will make the world around us seem much stranger than any of us can imagine.'*

But does a better understanding have to make things weirder? String theory hinted that space and time are approximate ideas, to be replaced by deeper concepts. This is really 'where we've got to' in physics. String theory left the impression - a misleading impression - that the small-scale structure of space must be very complicated multi-dimensional patterns, all folded up together, which somehow lead to the physics in our large-scale world.

So hardly anyone bothered to look for the simple geometry that nonetheless might be there. (It is there, this has been proved.) That's why so few dared to tackle the conceptual side: it seems unscalable. People decide that searching through the mathematics for a way forward is a better bet.

But although it seems daunting, a simple setup can have complicated results, far weirder than their source. So if you can only see what emerges, it might seem that something strange and complicated is going on. But we shouldn't assume that it's hard before we start. If you think you won't find something, you probably won't. And the universe has an odd habit of turning out to be unexpectedly simple. I decided in the early '90s that this simplicity would go on into the next paradigm, and then started searching. Victor Weisskopf was greatly respected - he always talked about the importance of the conceptual side, and used to say 'Search for simplicity'.

A loose illustration for this point about the source involves a spinning skater. It's not a good analogy, but you'll get the idea, and that's all it's intended to do. Instead of bringing in her arms to illustrate a well-known principle, as she normally does, today the skater is holding a handheld firework, and waving it slowly and evenly up and down as she spins.

That's a simple setup, with the smoke from the firework moving outwards in all directions. But suppose the lines of smoke are all we can see. The patterns that go curving around might look surprisingly complicated. They go up and down, also around, also outwards. Perhaps there's enough combined motion to make it hard to tell the cause. It might seem that whatever is generating these patterns must also be something very complicated. And it's difficult to work backwards, as we need to in physics, and try to guess what the simplest possible setup that could create that pattern might be.

In the same sort of way, the concepts for a new understanding of space and time could be far simpler than people expect. And as Einstein and Wheeler believed, there might be some truly simple mechanism - but an unexpected one - sitting there underneath it all.

Part 6. Is there a flaw in our present picture?

23. Three logical steps

The twentieth century stretched out before Einstein and his colleagues when special relativity was discovered at the beginning of it. But after a while they realised, to their amazement, that the physics was telling them all its events were already mapped out and unalterable.

There were three logical steps that led to block time. The result left no room for motion, no room for matter even to have a speed. Events don't 'happen', they just exist in a motionless four-dimensional fabric of points in space and time. People can't shape what happens at all, even if they seem to shape it. This looks a bit like what some people call 'fate' - and some simply accepted it, rather in the way that some simply accept fate. Without any doubt at all, some saw block time as a depiction of fate. So how did we get to that weird, frozen picture in the first place?

The starting point was simultaneity across a distance. Before the 20th century there was Newton's view of time, with the kind of simultaneity we expect from the universe. At a given moment, a huge number of events across the universe all happen at the same time. Clocks tick evenly everywhere, staying in sync if they're well-made, and all events are linked up across space in an obvious, intuitive way.

But when special relativity arrived, it led to simultaneity becoming relative instead of absolute. The timing of events some distance away depended on how one is moving (and this is not about light travel times). To one observer a distant event is happening right now, but to another, moving differently, it might already be in the past, or still in the future. So across a distance, where the present moment lands depends on how one is moving.

The new set of rules was weird, and it was no longer clear what simultaneity actually *meant*, but the system had an internal logic to it nonetheless. And it seemed possible to relate 'now' moments in this way (even though we found out later that they're not meant to exist). At the time, making this change to our picture seemed reasonable, and harmless enough - particularly before we found out where it would lead us.

But to construct this relative version of simultaneity, assumptions about time had to be made. We assumed it was like the other three dimensions in being a large-scale one, *spread out across space*. When the underlying assumptions were made, no other kind of dimension was thought to exist, as Kaluza-Klein theory was still twenty years away, with its idea that a dimension can be very small and 'curled up'. In the early years there was far less reason to question those assumptions than there is now.

So of the three logical steps, step one involved assuming a specific, limited simultaneity across space. Step two was a conclusion that came directly out of step one - *a single event can be in the past for one observer, but in the future for another*.

And that, when people looked at it closely, was a bombshell. It transformed our species' view of time. It did so because of the following reasoning: if the event was in the past for one observer, it was inevitable. But if for the other observer the same event was still to come, and in the future, that meant a future event was inevitable.

That second step seemed to show that past and future are interchangeable depending on ones viewpoint. So it seemed that there was no real difference between them - except in our perception of them.

Step three involved standing back a little, and applying step two everywhere. Because this inevitability can apply to any event, it would therefore apply to all events. As any future event approached, there was always going to be a possible viewpoint (for some hypothetical observer, moving in a particular way) in which it had already happened. That meant block time, where all events are inevitable, and the entire future is pre-decided.

And this applied not only to future events: all past events were inevitable too. That conclusion might seem less drastic, but according to the standard view, suddenly from 1966 it had been 'proved' that it was utterly inevitable that our species evolved and developed *just as it did*, and went through the exact history it went through. That meant right down to the smallest detail - including a vast list of comic, tragic and very boring events that were always going to happen. Physicists quietly accepted this, but the public didn't - they didn't know about it. No-one breathed a word.

This inevitable, frozen picture has had a devastating effect on both physics and philosophy, and for half a century we've been unable to go on anywhere from it. The word 'block' was appropriate - the concept was like a roadblock, stopping all progress. It's still the standard view, and although people have found ways to ignore it, no-one has found a way out of it.

But without motion through time, we lose a large part of our science. Physics describes the world in terms of motion, action, cause and effect - all of which need time and change to drive them. Block time seemed to threaten the very heart of our picture of reality, by removing the central engine that drives it along.

Block time implied that motion through time is unreal, in what is effectively a fourth step. It went like this: if past and future really are no different from each other, then how can you have a moment called 'now' moving along and defining a difference between them? There seemed no way that could be possible, so many decided that the flow of time we experience is an illusion, *even with the measurable variations to its rate*. There are also places where the time term 'disappears' from the mathematics, suggesting to some that there might be some complicated illusion at work.

But this meant an absolutely crucial component of what we observe in the world around us was now missing from our picture. It seemed beyond the scope of present day science to pin it down, which greatly limited science's ability to describe the world.

And philosophers have had a terrible time with it. When they try to define how something or someone exists in relation to its or his or her history, they divide loosely into two groups - one group that ignores block time a bit more, another group that ignores it a bit less. Both groups are forced to ignore it to some extent, or philosophical nonsense is never far away, in which there's absolutely no connection between an object at one moment and the same object at the next moment.

And in physics, the issue of a flow of time has been gently booted out of the field. In the 20th century it seemed to many physicists that its origin might be psychological, and somewhere deep - creatures would have evolved with this illusion built in from the start. That would make it very hard to find.

But because the measured time rate changes are accurate to billionths of a second, that part is often seen as a second illusion, distorting the first one, like a four-dimensional perspective type effect. But two interacting illusions is a contrived explanation. There's not much that *can't* be loosely explained away, if one is allowed to invent two undefined illusions. (Try it out, it can be a fun game.) So instead there's likely to be a flaw in our thinking.

24. A conceptual piece in the wrong place?

It has often been pointed out that a single false idea, in any view of anything, can be like a spanner in the works, causing problems in many areas. There's a great well-known remark, often attributed to Mark Twain, that goes: *'What gets us into trouble is not what we don't know. It's what we know for sure that just ain't so'*.

This is such an exceptionally good point that the credit for it is well travelled. It has been attributed to Josh Billings, who probably said it first, but also to Mark Twain, Will Rogers, Artemus Ward, and more recently to two baseball players, Satchel Paige and Yogi Berra.

Another well-known saying puts it in a different way: *'The greatest enemy of knowledge is not ignorance. It is the illusion of knowledge'*. Although widely attributed to Stephen Hawking, the source is the historian Daniel J Boorstin, who made the point many times in his books, showed how important it was, and how much it altered the history of science. A book of his quotations had Boorstin's version as the title, but he credits Josh Billings for the original 19th century version.

The wordings of this saying vary, but for physics, they lead to the same basic idea. It's far better to have completed less of a jigsaw, but with all the pieces definitely in the right places, than to do more, but with uncertain areas. One false assumption, and we could be delayed for a year, a decade, or a century. Many physicists are careful in this kind of way, but some, it has to be said, tend to cram all the pieces into the gaps in the jigsaw, and push them down as hard as they can.

So given the assumptions that went in, and the contradictions that came out, it's clear that a false assumption might be causing the problems with time. If so, its arrival (probably in 1908) would be unsurprising, as it'd be something very hard to identify. But if it's still in there, the question is where - it could affect the whole picture, so there'd be a need to find it. It's worth examining the Rietdijk-Putnam argument, the three steps that led to block time. They seem to indicate that our motion through time is rooted in nothing but our perception. Underneath that there's also Minkowski's work: the set of ideas from which that whole bit of reasoning arose.

But deeper still, there's another set of assumptions underlying those levels. They're about things like frames of reference, or how we take objects to be moving. We may have got them right, but some were untestable, and some were based partly on intuition.

And more generally, seeing it as a jigsaw, because a piece of the puzzle may be in the wrong place, we should look at what's in place, and see which areas we *know* to be right, and which areas we only think are right. At present, what's on the table is a confusing mixture of both, and a lot of people don't know which are which. But if we're looking for a piece in the wrong place, it's essential to separate them off.

Special relativity itself, as Einstein published it in 1905, has been beautifully confirmed many times - it simply has to be right. But it just 'does what it says on the tin', without giving much idea of what's going on. Except for Einstein's vital interpretative step, that is, which was to say time changes its speed in different situations, from different viewpoints.

Special relativity also strongly suggests the dimensions are involved, which is what led Minkowski to spacetime. He showed that some things work well in the mathematics if you treat time as a dimension.

Spacetime and special relativity are very often taken as one chunk of jigsaw, but they're actually two chunks placed together, that may or may not fit like that. Only one has been confirmed. To test Minkowski's assumptions about time, further assumptions would have to be made about the experiment - it then gets more or less impossible to confirm anything.

And in the 21st century, because of the need to sort out time in order to get to quantum gravity, questioning spacetime has gone rapidly from a heretical outsider's position to an acceptable mainstream one - in ten years or so. This is partly the influence of Lee Smolin, who has broken the taboo, and made it acceptable to look for a major flaw in the existing picture.

In jigsaw strategy, if the picture isn't fitting together too well, you look at any uncertain pieces. In physics that means untested ideas. You try taking them out of their present positions, and moving them around. That's what people are doing now, so it's an exciting time.

25. Making room for new ideas

Classical physics was our picture of the world before the Michelson-Morely experiment. That single experiment, which answered a vital question, meant that whole view of the world had to go. In a few years it had gone, remaining only as an approximation of reality.

Newton had seen space and time as an unchanging backdrop, unconnected with the world of matter. And so had everyone else - this had been taken as *a priori* for centuries, before and after Newton carefully wrote it down. But from 1908 it was quickly replaced by four-dimensional geometry, in which space and time shifted around. This happened for good reasons.

But since then the path has led to contradictions. Progress is being held back, our theories disagree, and a further piece of baggage might have to go. Some are starting to let go of spacetime, but not special relativity itself. Only one of them has been tested, but it isn't only that.

Special relativity may be counterintuitive, hard to visualise, and downright weird in places. But it uses surprisingly few assumptions. Spacetime needs a lot. That's because it makes time a dimension like the others in some ways, different in other ways - and there's a long list of differences.

And being closely associated with the dimensions doesn't necessarily mean *being* a dimension. Some say George Martin was 'the fifth Beatle', but that's a phrase - he was certainly very important, but he wasn't one. So rather than trying to make George Martin a weird, different kind of Beatle, with a long list of differences from the other Beatles, we might do better to drop those assumptions, and try to understand him in other ways.

And losing spacetime does the same: it removes the untested assumptions that go with making time a dimension. It's not obvious that spacetime can be removed via Occam's razor - it has simplified things, both in the mathematics and conceptually. But Occam's razor is about *explanations*. There's not much of that either way - spacetime provides a little extra explanation, but at a huge extra cost in assumptions.

Another thing spacetime does resembles Nils Bohr's approach to quantum mechanics. It tells us to stop asking questions about the clues. A view known as 'the inseparability argument' holds, as Minkowski did in his famous 1908 speech, that space and time are part of one unity. But that approach already contains assumptions that could be false. It shuts down attempts to solve the puzzle, by plugging the holes in the jigsaw. Only if we admit it's incomplete will we leave room for a solution.

Spacetime also does a lot more assuming that explaining. But it facilitates the mathematics, and that it does in spades, in many places. This protected it for decades, but nowadays people are trying to replace it. Lee Smolin has been, with colleagues - a 2011 paper shows their attempts to describe the world in terms of 'phase space' and 'momentum space'. This is an attempt to move around the blockage, and on towards quantum gravity.

There are tens of ways of taking special relativity. The mathematics itself is very simple, but its application can be shifted around: it's like a Rubik's cube type puzzle, that can be moved into many different configurations. Still, most versions we know use the same kind of assumptions.

But because there's 'weirdness' in the theory - it's counterintuitive - there's room for false ideas about it. Having taken the weirdness onboard, we have to accept that it too can move around: the weirdness could be in a different place from where we usually put it. For instance, I found a lot of reason to think that the weirdness is in the transmitting medium, rather than in what it transmits. But in a general way, those *kind* of unknowns leave room for our assumptions to be very wrong.

I wondered if this could be turned into a principle, like Occam's razor, but it came out looking a bit obvious: 'With any system that's not understood, the more counterintuitive its behaviour, the less reliable untested assumptions about it are'. This is so general, it applies even when trying to diagnose and repair a car engine.

Spacetime is a set of untested ideas. So block time can't be depended on, but if the 'proof' that led to block time is wrong, it must contain a flaw, and there are only a few places it could be. So searching through the assumptions, we might find the fly in the ointment.

No-one has ever suggested a flaw in the reasoning that led to block time, as far as I know. If anyone did, it better be good, or it'd be pulled apart rather quickly. Later, in Chapter 43, I'll set out what I think that fly in the ointment was, and what it did to our mathematics. Until you see it, it looks impossible for anything to remove block time - according to my view that's why we've been blocked for so long.

Part 7. Is the future already decided?
Quantum theory v. block time

26. Opposite messages

Among the key issues surrounding time, there's the question of determinism and freewill. There has always been genuine confusion among thinkers about what ultimately causes the pattern of events in the world, how predictable it is, how much we shape it ourselves by our actions. Into this confusion came relativity and soon afterwards quantum theory, which pulled in two opposite directions on the question. They've muddied the waters further, but frankly they were quite muddy already.

The view that came out of Newton's work was of matter obeying clear and simple laws, in a world that seemed entirely predictable and deterministic. But this left questions about where human decisions fit in: people wondered whether that left anyone responsible for anything. This was about the mind and its relationship to the world, so it then went off into deep water.

Relativity appeared, suggesting that all events might be pre-decided in some other way, not because of the predictability of events due to physical laws, but because time was already laid out like a dimension. Then, twenty years later, quantum theory showed that small-scale events are not predictable in the way Newton's laws had suggested anyway. There's an utterly random element, and it doesn't go away with more information.

So the picture of a reliable, predictable world that came from the old physics was gone - the place where it used to sit was now sandwiched in the middle. Relativity has made things look even more pre-decided at some level, while quantum theory has made them look far less so.

That's where we still are now - the result is surprising. It's well known that relativity and quantum theory are incompatible. It's generally thought that we'll end up with a theory encompassing both, of quantum gravity. Although the differences have increasingly been seen as a problem, some still see it as the need for a bit of minor tweaking here or there.

And yet the difference about the future is non-negotiable. It's not the kind of thing that a minor adjustment will fix: the small-scale and large-scale worlds are sending us opposite messages about a major, fundamental question. But the issue hardly enters the mathematics, and many physicists ignore things that don't. So the point either gets ignored, or 'thrown to the philosophers', or simply filed under 'Things we don't understand about time'.

27. A mismatch between scales

If events are truly random at a small scale, as we find them to be in quantum mechanics, this unpredictability must also affect large-scale events. It would make no sense to imagine that the future is undecided for particles, but pre-decided for us. Further up the scale where we live, every large-scale event is made up of many small events.

If what will happen is undecided with small events, connections across scales would make the large-scale world's future unfixed as well. It can be shown (as Schrödinger did with his cat) that there can in principle be connections of that kind. And with connections across scales, a little goes a long way - it's a bit like the butterfly effect. An open future at a small scale would leak out: a slight change here, another there, and further down the line the large-scale world is very different.

So the future can't be pre-decided, and the puzzle isn't removed by different rules applying at different scales. But this still looks like another mismatch on a long list of mismatches between two scales. We use relativity for the large-scale world, quantum theory for the small-scale one. The two theories don't match up, nor do the scales. In fact, in standard physics, it's hard to see how the large-scale world emerges from the small-scale one at all. If we apply our current physics to the smallest scale, we find total chaos, and we can't see how the neatly ordered large-scale world emerges from that.

So the mismatch about the future is part of a wider, more general problem. It means either we don't understand how the two scales connect up yet, as many assume, or worse, one of our ideas is wrong. The contradiction about the future, in which it's impossible for both pictures to be right, looks like an indicator that it's the latter - that one of our ideas is wrong.

Another strong indicator that an idea is wrong is one of the few places where the two scales should meet up in our calculations: a number for the energy density of the vacuum. The amount of energy contained in one cubic meter of empty space figures in work on the large-scale universe. But it can also be calculated from the small-scale one, via quantum field theory. The prediction from QFT has been measured and more or less confirmed, with the Casimir effect.

Unfortunately, the number we confirmed from the small-scale world is 10^{120} times larger than the result from the large-scale world. The small world gives a big number, the big world gives a small one. The factor difference between them is a 1 followed by 120 zeros, but they're meant to be the same number. Known as the vacuum catastrophe, the problem has been described as the worst prediction in history. So far there are more or less no explanations for it on the table, but later in this book a viable one is set out.

So standing back, the crack in our picture splits scales. We find different rules down there, and have spent a century trying to find out how and where they change over to what happens up here.

Whatever the answer, time is near the centre of all the troubles, and is both where we come unstuck, and also where we might find a way forward. The time issues have been a problem for physics and philosophy alike. It seems that these questions constantly lead us into deep water, but the hope is that 21st century physics will find a new approach that removes the difficulties. So it's worth looking at the two alternative views of the world, given that one of them must be wrong.

28. Future fixed v. future unfixed

Looking at the implied fixed future of quantum theory, you see how hard to avoid it is. Bell's theorem ruled out 'hidden variables' in many places, making the randomness almost inevitable. For decades, quantum theory has quietly pointed towards an open future. Although it's a taboo, that makes block time and spacetime both look wrong. A slight element of randomness *anywhere in quantum theory* does it, and the large majority see events as unpredictable. In that poll I talked about in Book I, 'There is a hidden determinism' got 0% of the vote.

I should explain exactly why the randomness means an undecided future. If what happens in an event is truly random, then until the selection is made, it could be a lot of different things. If so, until that moment, what will happen doesn't yet exist. So nor does the future - that's the key point, and what the randomness does.

But spacetime tells us the exact opposite. It implies a future already laid out, and means that what to us is the present moment is no more real than some future or past moment. The future, according to spacetime, is not just some pale shadow of an outline, hinting at what will eventually fill that space. It exists 'now', throughout history, and is as real as anything else.

So it's a very specific conundrum, in which *diametric opposites* are implied by our two central theories, almost as if the universe was teasing us. And this puzzle, which in the past was taken to be about theology, can of course be taken to be about physics. But physics students are rarely even told about the problem - most don't know about it at all.

It's worth pointing out that both of these two scenarios are alien to us. And oddly enough, although they're opposites, both are alien to us for the same reason: they both leave cause and effect behind.

A truly random selection out of many possibilities is not like the cause and effect we're used to, on which we base our science. But then nor is an entire history that's pre-decided - *both* leave this vital principle out, in which events arise as the result of other earlier events.

The randomness is far more acceptable and likely. Unlike spacetime, it leaves plenty of room for cause and effect. The odds can be calculated for what will happen, and the randomness is kept to a specific, limited area. And the same randomness leaves room for freewill to be at work, as our experience and observations tell us it is. Quantum theory shows a world with room for us to be contributing to what happens. That's important, as we struggle to arrive at a picture that truly fits the world we find around us.

Block time, by contrast, leaves no room for cause and effect, or freewill, to exist anywhere. To get either, you have to split the picture into two separate levels. People always used to assume we could, so they didn't think anything was amiss, and would push the problem to one side. But we're simply not in a position to do that.

So if we have to choose - and we do - then my bet is on quantum theory, not spacetime. But in looking at which of the two is more likely, there's one point that beats all others. It's that we sit and watch the randomness of quantum mechanics at work in laboratories. The fact is, we've found the world to be that way. But the pre-decided future from spacetime has no evidence at all. It arises only on paper - in one of the few untested areas of a theory so well tested in other areas, that people take it to have been confirmed.

29. Comparing the routes

But there are two branches of this conundrum, not one, and the flaw could be on either branch. We reach these contradictory conclusions on the future, so it seems *somewhere* in the reasoning, there must be an error. Perhaps the mistake is in our general way of seeing one of the theories. But whether it's a background assumption or a foreground one, hiding nearby, somewhere not too far away, there must be a false idea.

I won't talk about the issues of quantum theory, as Book I is about that. But people have been trying to remove the randomness for ninety years. Bell's theorem made the randomness likely to be fundamental, and although it's a taboo, if it is, block time is false.

This crack in our picture is so rarely talked about that when an essay paper of mine making the point went up on a site in 2012, I didn't know how people would react. In the months that followed, many comments on the site, which is a community of physicists, showed support - there was no criticism for the premiss. This is not politeness, these people sometimes argue like mad. I'm not saying everyone agreed with the points I made, I'm absolutely sure some didn't. But it's very difficult actually to *argue* that the randomness and block time can co-exist, even if one thinks it.

Anyway, for the rational view: truly random events means no block time. So it seems there's a glitch somewhere, and the two routes that led to the two contradictory pictures don't look like a 50-50. The quantum mechanics route looks far more reliable. And the spacetime route? Well, we've been changing our view of spacetime anyway.

30. Room for error

A general feature of what I call the unreliable route - the spacetime route - is room for error. This can be seen clearly in places. Firstly, the landscape from which block time emerged *included* the timeline motion. Some philosophers have pointed out that the 'proof' of block time came out of concepts such as a present moment, which it later apparently shows not to exist.

So block time came from a set of ideas, timeline motion included, that it later shows to be partly wrong. In doing so it attacks its own foundations, like an electric drill that risks cutting through its own power cable. But most people who have taken this into account still think it's right anyway. I'm not arguing that spacetime doesn't lead to block time - it does. But with elements of this kind in there, there's more general room for error.

It's also worth noting that block time depends on the rules of simultaneity at a distance being *exactly* right, and the rules about time for reference frames. A single minor difference could make our whole present view of time wrong. That means a difference to a dimension thought to have unique differences from the other dimensions already.

And that brings us to the absolutely number one disadvantage of that whole approach. It's this uniqueness of the time dimension. The most worrying of Minkowski's implied assumptions was the last on the list, which effectively goes *'...and this is the entire list of differences between time and the other dimensions, there are no others'*. We simply can't assume that. Given the problems with time, It's very likely that we haven't found all the differences. As we've accepted some, there clearly could be others.

And nowadays string theory has led many to see what we call space and time as approximate concepts, which emerge from deeper ones. These concepts are thought to be unknown - I'll suggest some later on. So time is taken yet another step away, increasing the room for error further.

So those are the two routes: the reliable one, and the unreliable one. Going by rational thinking, there has to be a flaw on one of them. And although we don't know what it is, we can see its result: *our overall picture is inconsistent*. The timeline motion is needed in places, but banned in others. But when we try to remove it, we take out a piece so fundamental, like a structural beam from a house, that the whole thing tends to collapse. And that puts at risk even the arguments that carried us to that point in the first place.

But because spacetime looks right, on the face of it anyway, there have been enormous and endless efforts to appease and accommodate it. Much of the time-related physics of the 20th century is a history of attempts to make the timeline motion go away, and much of the confusion arises because it won't. But another way out of the main contradiction - that this motion seems both to exist and not to exist - is the idea that it somehow does exist.

And another way of seeing block time is that we're getting nonsense out of a picture that's mistaken somewhere. So in one way of looking at it, block time is like a dragon that has been giving us a very hard time, and needs slaying - but that's easier said than done. Either way, we've reached a point where we can't just sit back and accept the landscape, as we very often have done up until now.

31. Very superstitious?

But the landscape was often accepted for another reason: to some it looked as if these two levels work independently. At one level, history is already laid out and pre-decided, but at an 'everyday' level the universe is unpredictable and the future is unformed. The everyday level has motion through time, the other doesn't. Of these two windows on the world, to some the first looks like a 'view of the gods', and the everyday level looks like humankind's view.

And because both theories have been confirmed, it seemed that both might somehow be true at the same time.

This lands near an unscientific version of time. It has freewill, but everything was still always going to happen as it does. The implication is 'well, it might just be like that'. But you can't do that - well, you can't drag that into physics. That's something else: metaphysics (using the term loosely), non-scientific philosophy, or theology.

The key point is, if that were true, we wouldn't expect our physical theories, or experimental results, to tell us so, because it wouldn't be a physical aspect of the world. So it wouldn't have any bearing on these questions about our physical theories that we're looking at here.

In physics the picture has to be self-consistent. We look at all kinds of weird ideas in physics, but they still have to hang together in a certain rational kind of way, and they're always about the physical world. And in this rational way of seeing things, random events means an undecided future.

But many physicists have this split-level picture implied in their view. Traces of it appear in conversations and email exchanges. But it isn't talked about, because if you take a look, you find it isn't science. In science you can't take a direct contradiction, and put the two elements at separate levels, without saying anything about what these levels are, or how they can co-exist - and how this contradictory scenario is possible. And when you get right down to it, the future either exists or it doesn't.

But to some this two level picture looked like it might perhaps be the case. It looked like something far beyond our scope, perhaps even supernatural, and it seemed we had reached a boundary at the edge of physics. So people tip-toed away, and left the area well alone.

And the same thing happened elsewhere - in quantum theory. Ideas that are unscientific get in. I'm not saying that this kind of view automatically has to be wrong. I'm just saying if it isn't science, it should be somewhere else.

To put it in a rather odd way, suppose someone was told that there's an idea believed by millions in the 21st century. It can't be proved or disproved, or tested, but it means that all of human history - *in every detail* - was inevitable from the start. They'd naturally guess it was something coming from religion. But that idea is Minkowski spacetime.

The spacetime picture started in the early 20th century, and the implied fixed future, and the inevitability of human history, slipped into our physics more easily at that point than it would have done later. It walked straight past the security guards, and was waved on through all the checkpoints of skepticism that such an idea would encounter nowadays. The generally assumed view of the world that preceded the 20th century was still deeply ingrained.

And in 1908 quantum mechanics hadn't arrived yet, with its random small-scale events, so the most glaring contradiction was not yet in front of us. And the *proof* that spacetime leads to the inevitability of all events hadn't arrived either. So spacetime slipped into our physics effortlessly, and without filling in a form, at the time.

But nowadays we question everything, that's what we're like! Whatever one believes about the world, and we're very open to all kinds of ideas, we don't just *assume* that if we find some weirdness in physics, there's some religion-related background picture behind it. Spacetime is certainly science, but one aspect of it that can appear isn't: the unscientific idea that the future is both entirely fixed, and also entirely unfixed.

Of course, many are not superstitious about this point at all. Instead they've just looked at the mathematical picture - and perhaps not very much else - and decided that quantum mechanics can fit with block time. But elsewhere, superstition has been an element of this.

The same happened with quantum mechanics, where a lot of physicists and philosophers accepted the 'woo-woo', without understanding it. To be fair, it certainly seemed to be true. But the main reason that elements of woo-woo remained in these two puzzles was that where an area of science looks like it goes outside science, people would just avoid it, and work around it. So the two puzzles just sat there, and progress on them was slow.

Incidentally, I'm not commenting on the 'big questions' in saying that there's a need to remove some unscientific ideas from physics. I'm just saying how I think science should proceed: as it always has, sticking to ideas that are clear and specific enough to be pinned down.

This book aims to show that these unscientific elements, found hiding deep within our inherited view of time, which kept people away from questioning things for so long, were false. It's hard to slay the dragon of block time, and show Minkowski to have been wrong, as people suspect he was nowadays. But we'll slay it in Chapter 43, and I'll try to show the answer to the question 'is the future already decided?' is 'No not yet! It's still wide open….'

Part 8. Which avenue might we unblock?

32. The trouble with mechanisms

Every day we move along the timeline. Some think this timeline motion is an illusion, but if so, it's not a hazy one. Measurements of the way it varies its speed locally have been getting increasingly accurate for seventy years. One measurement made at NIST in 2010 was done on an object moving as slowly as ten meters per second, creating a time rate difference so small that some thought we'd never reach that level of accuracy.

The measurement was from one of the world's most accurate clocks. Even at just 10 m/sec, events onboard the object were running slower. Each second was longer by about half a million billionth of a second - just as expected, via the simple mathematics of special relativity.

People used to think this effect is just about how the world is. But nowadays many think there's more to be discovered, partly because of the shift in our view of spacetime. So an open mind is needed.

And on the face of it, it looks like some precise physical mechanism is at work underneath what we observe. But the trouble with that idea is, any physical mechanism needs time if it's going to work. So it's hard to see how any such mechanism can also generate time. If it's the kind of mechanism that might create the local time rate differences, it would need time to function. So it seems time would have to exist already.

This has been a problem for many attempts to explain the puzzle by bringing in a mechanism for the time rate variations. It's hard to get creative with a mechanism that adjusts the time rate locally, because it's hard to invent a mechanism that could work at all.

This kind of restriction also affects other attempts to get out of the problem - some nowadays are trying to make causality, for instance, run deeper than time, and with time as emergent from that. But causality (cause and effect) needs time to exist already, because there's already a sequence of events implied. Time runs so deep, and is so essential to physics, that it's hard to 'get underneath it', and do any physics without it - in order to generate time in the first place.

So some expect a different kind of explanation, perhaps involving weirdness of the time dimension. But that hasn't worked either. As I'll try to show later, and as some good physicists have started saying this century, when looked at closely, it lacks self-consistency.

So - drawing together all the points in the book up to here, and looking at the whole puzzle - there are several different avenues that lead off from where we're standing. But each of them, without exception, turns out to be blocked somehow, in one way or another.

Thermodynamics, which we'll get to, is not one of these avenues, as it's used in attempts to explain something else (the direction of time, but if taken as an avenue about motion through time, it's also blocked). The result is, it's far from obvious what *kind* of solution we should be looking for. It sometimes helps to stand back and look at the different avenues. We might be able to reduce the possibilities before even starting the search. And looking at these blocked avenues, one of them - and only one, it seems to me - has the kind of blockage that might be removed.

33. More than a minor adjustment

But it looks like the adjustment that's needed will be a major one. Attempts to make a minor adjustment, and there have been *very* many recently, don't seem to be enough. That means we may be near to a paradigm shift, which is exciting, even though many are still trying to fix up the old picture.

George F R Ellis, of the University of Cape Town, is a well-known and highly respected physicist - early on he worked with Stephen Hawking. He argues strongly that there's a real flow of time, and has neatly shown other views of time not to work.

I was very glad to have an exchange with him in 2012, in which we found we agreed on some key points. We both see a conflict between quantum theory and relativity about whether the future exists. His recent approach, called the emerging block universe (EBU), involves what's really a minor adjustment to both spacetime and block time - keeping a form of spacetime, but not as Minkowski had it. The idea is that the block crystallises steadily, behind 'now' moments that move through it. This is an attempt to leave the future open, while the past is already frozen into the block.

To me, this hasn't washed out all of the problems, and a central one remains, which is that what is past and what is future are still different for different observers, and still depend on how one is moving, as they do in the standard picture. And that dependence on the viewpoint, for past and future (which we'll get to soon), is what made time look like a perception-based illusion in the first place, causing the problem. It's very good that George Ellis has been grappling with these questions, along with the rest of us. His support for the idea of a flow of time has been a real boost for that general view.

But because of the deep flaws, rather than a smallish tweak, an entirely new interpretation might be needed. Searching through existing physics, as many are doing now, for some small effect we overlooked - which is causing time - isn't enough. A mechanism like that needs time already underneath it, and it should be admitted. But that's a paradigm shift. So instead, people seem to hope that at a small enough scale, that kind of problem gets smoothed into the background, and just goes away.

Whatever the answer, motion through time doesn't fit our view of the world. So we work with it without knowing what it is, as people used to work with electricity. And the discussion goes on: many try to make a view of time out of the jigsaw pieces we have. Some talk as if the jigsaw is largely finished, but showing a picture that's hard to understand. It's a picture with good internal logic in places, hinting at weird but logical four-dimensional geometry, and it looks like it might well be a real structure.

But apart from the contradictions, there's still a lot we can't understand or describe. So it seems premature to assume we're going in the right direction. We've been stuck on this for a century now, and it may take a sizeable lateral jump to go on anywhere from here.

There's a good indicator that physicists are trying to fit the world into a false description. It's when attempts to fit time into the picture shift around. If our approach often needs resetting, and refuses to 'settle down' - over decades - we may well be on the wrong track.

And the ideas do shift around. There have been major U-turns in how we see time, and each shows how weak the preceding idea really was. One is the switch from seeing motion through time as an illusion, to seeing it as real, but emergent. Another is the switch from seeing spacetime as fundamental, to seeing it as emergent. Making something emergent - particularly from the unknown - is like when a painter smears the paint slightly: it blurs something into the picture that was a bit too visible. And as we learn more, we find that our old fixes, which we believed at the time, didn't really work.

34. An incomplete picture

Some physicists nowadays are trying to change the underlying landscape by 'shuffling the principles'. They assume we have all the pieces of the jigsaw in front of us, and only need to rearrange them correctly. To do that, some try to adjust the relationships between the underlying concepts, so things that were seen as fundamental become emergent, and vice versa.

So for instance, causality, or dimensionality, might become concepts that are more superficial, which emerge out of more fundamental ones. On the other hand, causality can be taken to be fundamental, and some try to get time to emerge out of that. These approaches sometimes lead to highly complicated physics - for instance, people have been trying out the idea that the number of dimensions of the universe isn't a whole number.

To me, a common mistake is assuming we already have all the pieces of the puzzle. I've always thought the answer will be simple - comparatively simple. But for that to be the case, there'd probably have to be *missing conceptual pieces*. So I think we should be searching for them. For one thing, a look at the general hole in the jigsaw, and a close look at the loose pieces we have, shows we don't have enough pieces to fill it. Finding new pieces might allow a simple answer. The list of clues on page 91 strongly suggests that there are missing pieces.

The reason I think the answer will be simple is partly the simplicity of a lot of fundamental physics, which many have pointed out. But more specifically, it's because of one thing - the stunning simplicity of special relativity itself. To me this suggests some truly simple setup involving the dimensions - and it implies, in the mathematics, a basic whole number of them. So it means that some are looking in unnecessarily complicated places.

John Wheeler also expected a simple answer. He said *"How can physics live up to its true greatness except by a new revolution in outlook which dwarfs all its past revolutions? And when it comes, will we not say to each other, 'Oh, how beautiful and simple it all is! How could we ever have missed it so long!' "*.

For what he's talking about actually to arrive, you need missing conceptual pieces. But surprisingly few mention the idea that we might be looking at an incomplete picture. To search for missing pieces one has to enter an area full of unknowns, but the deepest cracks in our picture are the very places where the best clues are to be found. If we really do have the wrong picture of the world, then the deepest contradictions in it should give clues about what the right one will look like. Motion through time made a good starting point for me twenty five years ago, for that reason.

There's an excellent quote, often attributed to Isaac Asimov, which goes: *The most exciting phrase to hear in science, the one that heralds new discoveries, is not "Eureka!" (I found it!) but "That's funny…".*

John Wheeler also knew that areas we don't understand is where to look. He said: '*in any field, find the strangest thing and then explore it*'. Although he didn't say the best clues are to be found there, I don't think he was exploring strange areas of physics just for the sake of it, in the way some people watch 'The Twilight Zone'. He wasn't being weird or kinky - he was there because that's where you find things out.

Part 9. Removing emergent time

35. Why the nature of the dimension can't solve the puzzle

If emergent time could be shown to be false, that quite simply removes the present view of time. It's dependent on emergent time, in every version. But before showing a way to do that, I need to remove a view I call 'split-level' time, which is absolutely everywhere. It's implied just beneath the surface of the standard view, in many places.

At the heart of many people's view of time, there's the assumption that both the two levels we find must somehow co-exist within the time dimension. If so, perhaps we don't have to explain these weird clues at all.

Of the two levels, one is frozen (as in block time), and the other has motion happening on it (as in the motion through time we seem to observe). And at first glance, it does look like the two levels might co-exist somehow. Perhaps the time dimension is weird enough for this two-tier structure to exist within it, with both levels part of its nature. It's a defeatist view: not a million miles from 'That's just the way it is'. But many quietly believe it, even though we have to build the time dimension up into an intricate phenomenon - unique, complex, and not like the other dimensions, to get there.

But the thing is, it still doesn't work. On close examination, only one of these two levels can be real. For now it might be either, but they can't both exist. I've mentioned the general contradiction about the future - there's a more specific version of it. It can be used to show that the time dimension doesn't have two levels within it. The reasoning goes: *'if it did, these two levels would disagree about whether the future exists, so that's impossible'*.

At the frozen level of block time, what to us is the future 'already' exists. It's just as real as any other era, and unalterable. But at the everyday level, with motion through time, you have quantum mechanics in action. And that has a randomness that for eighty years has been seen as a fundamental element of the world, implying an unfixed future.

So these two levels of time can't both be real. They don't describe the same universe, so the contradiction is too deep for what we call physics to allow it. This conclusion is actually very helpful, in a confusing landscape. It leaves us only two possibilities to choose from, so it narrows things down. It's simple enough: of these two levels, one exists, the other doesn't.

And in fact, strict relativists, and uncompromising supporters of the standard view, think the same. They also think that only one of these two levels is real, but to them the real level is block time. So they often invoke illusions to deal with the other level, at which things can move.

But nowadays a growing number of people are starting to think it might be the other way round. What if motion through time is the real level, and block time is unreal? That could remove *all* the contradictions, and every problem. And for block time to be unreal, it might be a little easier. The search would not be for illusions, but for a false assumption in our view, hiding quietly in there somewhere. And block time would simply be wrong.

It's hard to see how on Earth that could be, but in the infamous Chapter 43, I'll show a way for it to be true. Even just a possible way means a lot, as no-one has been able to find any way out of it at all.

But it is possible in principle, as among the main chunks of our basic picture in physics, the *one untested area* is exactly what led to all the problems - the spacetime geometry. And that was a set of assumptions about time. To put it another way: what led to all the problems was a set of untested assumptions about something we don't understand.

36. The first domino

Minkowski spacetime is like the underground station Châtelet Les Halles, in the Paris Metro. Many lines run through it: everything converges there, and all is well and good if Châtelet is in working order. But if it has a problem, the whole system will have problems, and Minkowski spacetime has been greatly weakened recently (however, the Paris metro is fine).

Spacetime is a central part of our picture, and it now has pressure on it from a few directions. It's partly that our view of time depends on spacetime, and that view is failing - even in laboratory experiments.

But it's also that in attempts to reach a theory of quantum gravity, and fuse the two main theories, spacetime is right where they should be joined. That's the area of physics being operated on, if you like, in the major surgery that's being attempted on our picture at present. But in the attempts to quantise spacetime, we're finding that physics is blocked, and doesn't provide a way forward - our picture isn't coming together.

If it did come together, each of the two main theories would give support to the other. But if not, spacetime is where the weakness is. Spacetime is a vital part of both special relativity and general relativity, and it's under pressure in both: with SR it's about problems with time, and with GR it's about problems with quantum gravity.

In this book I've tried to show some of the new reasons to doubt spacetime, also some old ones. But those aren't enough to demolish our present view of time - though the recent changes arguably leave it teetering on the edge of a precipice, as without spacetime it has nowhere to go.

But to demolish our present view of time fully, if one were trying to - and I do think there's a need to, to get to the truth and move forward - one has to demolish *one single idea* that many think saves our present view. Without it the present view is dead, but there's one idea that remains. If people see it as still viable, they'll go on defending the current ideas.

That idea is called emergent time. Our present view is entirely dependent on it, as it's needed to explain what we observe around us. It includes illusion time, thermodynamics and entropy, and a huge range of other ideas, and is far from specific about what's going on. But *some idea involving emergence is unavoidably needed*, if block time is to have any chance of working at all. So if emergent time can be shown to fail, the whole house of cards falls. And I think it truly can be shown to fail - I've found a way that others seem not to have found, or have not talked about.

And emergent time, it turns out, is just the first domino. If that goes, it takes out three ideas in order: emergent time, block time, spacetime. This is the reverse order from how they appeared, in which each depended on the last. Spacetime came first. Then it became clear that if it was correct, block time had to be as well. That turned out to need emergent time if it was to work. So if emergent time fails, all three dominoes go, and the third is Minkowski spacetime, which many are questioning now anyway.

And then, from there, because spacetime goes into so many areas, it shows that rather than endlessly trying to fix up the present paradigm, as we have been for longer than many people realise, the paradigm has reached the end of its useful life, and needs replacing.

37. Time as a superficial effect

Around the turn of the century, many physicists gave up on the illusion idea, as it simply doesn't work in certain areas. So they took a step back, and went to a more general, less specific idea - that the flow of time, or the direction of time, somehow emerges at a superficial level.

Emergent time is a wider set of possibilities. It includes illusion time, because if some illusion was at work, it would be an emergent thing. Illusion time is just one example - emergent time is *any view of time* that takes the apparent flow to be a secondary, shallower effect, rather than a deep one.

So when illusion time was dropped by many in the late 20th century, in a way people were jumping back, and saying, 'well, we don't know what it is, but it must be something of this generic kind'.

There's nothing wrong with that, but if emergent time is false, that removes a lot of ideas. And real problems accompanied the step back that was taken. For instance, illusion time had a psychological aspect, and without it, things that suggested time is a perception-based effect are now unexplained again. Explaining them was the aim of the thinking on time in the 20th century, but in the 21st, deciding time is a *physical* superficial effect, not a psychological one, some old problems have returned.

One problem emergent time has is that without an underlying flow of time, the laws of physics are like an electrical gift without batteries included. They simply don't work. Take an equation for the speed of an orbiting object: the vis viva equation. It has been used for centuries, and is still used everywhere, including by NASA.

The equation describes any orbit, and gives you the speed at any point on it. It might be a planet on an elliptical orbit around a star, following a repeating pattern. As it travels away from the mass it slows down - then it lazily comes around, speeds back up on its way inwards, flips around the mass, and heads outwards again.

But unless the planet can move through time as it does this, the equation has a different meaning. What it provides is distance over time, at any point on the orbit - how much distance is travelled in how much time. It's about what happens when matter moves through time, and if matter can, the planet can orbit, and develop - perhaps growing plants on its surface, or going through geological changes, as planets do.

Equations like that have a broad application - they seem universal. All that's needed is a few numbers. Knowing them, you can get an object's speed in a huge range of situations. But without some sort of flow of time, the equation doesn't describe an orbit, but something else. Instead it describes a set of ordered, looping patterns within a motionless block.

If so, the time element of the speed given by the equation is about what this pattern looks like in one particular direction. It's about the distance along the time axis within the block. A look at that, and one finds the law now doesn't describe anything involving the interaction of matter with matter. Whether or not gravity is a force - in block time, forces don't pull, objects don't move, causes don't have effects, and physics doesn't work.

A great many physicists, if asked what the equation describes, would say that instead of patterns in a block, it describes what happens at a different level. A common reaction is to assume there are two levels that somehow co-exist. That's what we do. But in the previous chapter I've shown that this can't be the case, as the two levels contradict each other.

Once you realise these levels can't co-exist, the block universe looks like the pinball game found in phones and computers. The game loosely simulates Newton's laws of motion. Press a flipper and hit the ball, it moves away as if Newton's simple laws are at work, but they're not. Complicated software simulates them. No forces pull, no objects move. Block time, in the same sort of way, simulates Newton's laws, but nothing actually pulls or moves.

I'm not going to discuss what the implications would be of finding the pinball game without there being real laws of motion anywhere for it to be imitating - that gets a bit weird. But in the case of block time, it simply doesn't make sense if the simulation is the only thing that exists. There are good physical reasons linking these simple laws together, and they're part of what we call physics. Many of the laws are arrived at from working out how objects pull each other around, and interact. Like many physicists, I believe that forces pull and objects move.

What all this means is simply that our laws of motion need some sort of flow of time. This is often ignored, as it hardly affects the mathematical picture at all, which is a kind of shorthand version. Time is taken as far more like space than it is in the full picture - in reality. This allowed the mathematics to carry on without worrying about this kind of problem, until the late 20th century. But by then people were realising that although block time doesn't mess up the mathematical picture, it plays havoc with the conceptual one.

It does this in a long list of ways. And in fact, the clues from the conceptual picture are usually the most reliable ones we get. Mathematical clues can be ambiguous - the same mathematics can be describing many different things. The conceptual side is our real mainstay, and if it's telling us there's an error somewhere, it shouldn't be ignored.

38. Emergent time removed

And here we come to one of the key points. In our present view of the world, with block time, there's no flow of time. Nevertheless, physical laws such as laws of motion were pre-implied, and frozen into the block.

This leads to a rather devastating problem for *both* the illusion idea and the emergent time idea. Either says time is a superficial effect, but the trouble with that is, it means these laws were just sitting there in the block, implied in the ordering of the time slices there, as if they were waiting for something more superficial to come along and make them work, by running the slices in a sequence. And something did.

That looks contrived - what were the laws doing there, sitting in the block in this 'just add water' sort of way? With illusion or emergent time, we have to assume that some shallower effect just happened to arrive 'later', in a largely unconnected way. And when it did arrive, it was so very appropriate for the situation, that it allowed these deeper pre-implied laws to work, by running the slices in a sequence, or making them appear to run in a sequence.

That's an enormous coincidence. It's a bit like finding by chance, lying on the ground near each other, a metal lock and a key that just happens to fit it, but which was there for some entirely unconnected reason. So the idea that the time slices were run in a sequence by chance simply doesn't work. Because they have laws of physics implied in their ordering, their very nature means that they were, as you might say - 'born to run'.

To me this point knocks out emergent time, in all its forms. It quite simply shows block time to be wrong. Emergent time is the first domino: it weakens the whole present view of time. And lab experiments at the quantum scale (Part 16) have recently done the same.

As a brief footnote to this chapter, there's only one possible way out of the apparent large coincidence that I've identified, to show that emergent time doesn't work. Some would say it can be addressed with anthropic reasoning. That's what Anthony Aguirre said, in an email exchange I had with him about this point, which started as an exchange about time. He said he thought the point was a very good one, and we got into a discussion about it. His view was that it may be a question of which universes have the right components to produce observers at all.

But the 'coincidence' identified here is not the kind that can have a random, probabilistic explanation. It's about a close connection between two things that had no need to be connected. The reason this apparent 'coincidence' can't have an anthropic explanation, is that it doesn't look like a coincidence at all. Given how unsure we are about time (shown by the lack of consensus), it simply means that something else must be at work.

If you want, you can try using anthropic reasoning to explain things that look *exactly* as if something else is going on - but Occam's razor will make these attempts look futile every time. Instead, the far, far simpler explanation will inevitably involve throwing out what made this look like a large, unnecessary coincidence in the first place - emergent time.

39. Layers of emergence

There's a wider angle that can be taken on all of this. Our world is often seen as made up of a number of layers, with the 'earlier' ones more fundamental, and the ones laid down later emerging from the deeper layers, each of them more superficial than the one beneath it.

Just as an aside, where patterns emerge from other patterns, words like 'pre' as in 'pre-existing' don't mean earlier in time, they mean more fundamental, or deeper. It's about coming 'before' other things, in the way the layers that make our world are laid down. The analogy is used a lot, for instance, there's a journal called Prespacetime - its name is about finding a layer deeper than spacetime. I've sometimes used words like 'already' in the same way.

Returning to these layers, physics has a series of them, and from there they go off emerging onwards and upwards. Out of Planck scale physics emerges particle physics, out of particle physics emerges chemistry, out of chemistry emerges biology, and out of biology emerge things like comedy, jazz, pizza, art, sarcasm, the internal combustion engine, and so on.

The layers in physics make a puzzle: the challenge is to identify them, then arrange them in the right order. The error in our present view is the idea that time comes *after* the laws of physics. Instead, wherever the laws go (in the ordering of the layers) time has to come before them - and run deeper. But our present picture is telling us to put time 'after' the laws. Many physicists, rather than taking this to mean there's a major mistake somewhere, simply comply, and try to make time more superficial than the laws. But ideas like illusion time and emergent time are vague, and they turn out not even to do what they're intended to do.

Some of us now feel these problems with time have been pushed to one side or tip-toed around for too long, and are determined to take them head on. It seems a genuinely new idea is needed, as Lee Smolin said in the well-known quote from him in the introduction. And it should be one that gets to *all* the contradictions - any approach that only deals with some of them would tend to look like a wrong answer.

But for a solution to remove all the problems, something in our existing view would have to give. And the trouble is, there's not much that could. We're dependent on the assumptions that led to the problems, but it's impossible to prove them - and worse, it's impossible to disprove them.

So instead they'd have to be replaced by something else. That's the only way we're likely to get past them, if they contain an error. This seems reasonable, and it's all very well to say that. But the fact is, it's a lot easier to criticise the standard view than to suggest something that might provide some kind of replacement.

Part 10. Shifting sands

40. A response to a glaring contradiction

A century ago, another contradiction in physics was causing problems. It was even more central than the current one about whether the timeline motion is real, as it was about ordinary motion through space, of light and matter. So unlike with the contradiction that we have now, everyone knew it had to be taken head on.

Special relativity was the simple early part of Einstein's theory. It was about matter moving in relation to other matter. The word 'special' is because it's about the special case of motion in a straight line at a constant speed. It was invented to reconcile two conflicting pictures, and solve a problem that had become unavoidable near the end of the 19th century. First you had the idea that objects move in relation to each other in the expected way. But now you had the idea that light always moves at the same speed in relation to matter. This had been shown by experiment, and also came (particularly with hindsight), out of Maxwell's electricity equations.

Both couldn't be true. The strange way that light moves in relation to matter didn't fit with our picture of the way matter moves in relation to matter. To reconcile these two contradictory versions of the world, something had to give. Einstein, thinking out of the box, saw it had to be time. This had more than a hint of 'when you have eliminated the impossible, whatever remains, however improbable, must be the case'. Others at the time, such as Poincaré and Lorentz, were beginning to try things with time as well. It was the only thing left to move around.

The problem was about speeds, and a speed is distance over time. It turned out that if time could vary, the mismatch might be removed. The equations for how a moving object is seen in slow motion were arrived at by making time change in a way that exactly removed the inconsistencies. And that was what drove the equations - there wasn't much else.

Einstein's version involved an interpretative step, emphasising the variability of time. Lorentz had already got to very similar equations, although Einstein said he reached them independently, as in a footnote in a translation he and Lorentz did together later (the Lorentz transformations are officially called the Lorentz-Einstein transformations). But Einstein was largely reinterpreting existing mathematics, and what he achieved was in many ways a conceptual breakthrough, as he found a comparatively simple explanation that needed few assumptions. It's one of the many examples that show how important interpretations are.

Einstein put light first and matter second. If light's speed was constant from any viewpoint, everything else would have to adjust itself to fit with that. To reconcile the two pictures, he calmly took one of them, the familiar world of objects, space and time, and bent it until it fitted into the other picture. This might give an idea of Einstein as like a plumber, who when one pipe doesn't fit into another, gets a hammer and makes some adjustments until it does. Instead it was a true insight into how two worlds meet, and experiment has shown it to be exactly right mathematically. It was among other things a new way to translate between reference frames.

The normal way of seeing special relativity is that it makes alterations to the behaviour of space and time. But without the presence of matter this could never be observed in action. So until we have a complete picture, we could instead be looking at a set of adjustments to the behaviour of matter. This alternative way of looking at it is really no weirder than the standard version, though at first glance it might seem to make even less sense. Neither makes intuitive sense, but it's not about that anyway. Given that neither approach is complete, either might lead somewhere.

But the point I'm really making now is that revolutions in physics arise out of a need - necessity is the mother of invention. When we realise how deep a crack in our picture is, it becomes clear that we have to stop papering over it, and instead do something about it. And a good start is having an open mind, and being open to new possible ways forward.

41. A history of resolved contradictions

And talking of dealing with contradictions, past physics has surprisingly often arisen in response to contradictions between theories. Maxwell's equations were incompatible with Galileo's basic picture, which led to special relativity. But then that was incompatible with Newtonian gravity, which led to general relativity. Attempts to reconcile special relativity and quantum mechanics led to quantum field theory, but that was incompatible with general relativity, which led to the present attempts at a quantum gravity theory.

So as we struggle to make our picture fit together and be self-consistent, we resolve one problem, only to run straight into another. These inconsistencies are mathematical, and they arise because in any phase so far, the conceptual picture we've been working from is only a provisional one.

Provisional pictures, even the sketchy ones that usually accompany theories, have allowed us to move forward in a series of steps. They're only possible at all because the same mathematics often describes many different pictures - equivalence is everywhere.

But although these temporary pictures have helped, they're not perfect, so they leave contradictions behind them. (The current problems fusing our two main theories suggest general relativity is wrong, as people have been saying recently - it's now widely seen as a large-scale approximation.) If we kept on in this way, the pattern of moving forward via efforts to reconcile conflicting theories might go on indefinitely into the future, from paradigm to paradigm. That is, until we well and truly fix the underlying picture.

That's why it's interesting that the present contradiction about time is not in the mathematics, it's in the picture. And many see the mismatch between quantum theory and general relativity as also being in the picture - this might be why decades of trying to reconcile them via the mathematics hasn't got us there. We don't yet know what's behind these conceptual problems, but the time issues are a central part of them.

The fact that our problems are rooted in the underlying picture might appear depressing, but it's less so than it seems. It may be that we're coming of age, and the present need to repair the picture will do something unprecedented. Perhaps necessity will lead to invention, as it does, and so we'll finally fix the picture properly. To do that, we might use our newest conceptual images of the world, from Planck scale theories. String theory, right or wrong, has given us a new picture of the smallest scale to look at. So we may be approaching a deeper resolution than previous ones, and a major step forward.

42. Is there a crisis in the present paradigm?

The way of seeing scientific progress I've mentioned, as driven by a series of attempts to fix incompatibilities, fits closely with Thomas Kuhn's view, as set out in his hugely influential book 'The Structure of Scientific Revolutions'. It was written in the 1960s and has become part of science, and central to our understanding of how science develops.

Kuhn pointed out that scientific revolutions arise after a crisis in the existing paradigm. He says science alternates between periods of confidence, where the current paradigm is doing well, and periods of crisis and then revolution - where the paradigm starts to fail, and is replaced by another.

The present crisis in physics, if there is one (people talk about one - opinions are divided as to whether there is one), is above all about the failure of string theory, and other theories, to get to quantum gravity. Thirty years ago string theory looked exciting and hopeful. But the mathematics was very new, and as work went on it divided into an enormous number of alternative theories, 10^{500} or more.

Those who say there's no crisis are often enthusiastic about string theory, and believe it will get us there in the end. But it has been largely untestable, and many now want to see a return to empirical physics, which we test as we go along. So the lack of confidence is partly about theory having gone where it's hard for experiment to follow. But it's also about theory having got stuck, without having achieved its goal.

String theory has been very polarising, leaving us in two camps. There's also disagreement on whether there's a crisis at all. It's arguable that as physics is a communication exercise, with published papers and discussion, if enough physicists think there's a crisis, there is one. If so, then it's partly about the incompatibility of the two main theories. There has been a failure to make our picture hang together, and bridge a deep crack in it. String theory is just one possible fix for that.

The time issues, although some try to deny their importance, are right near the heart of this division. And both the problems with quantising gravity, and those relating to time, are about Minkowski spacetime.

There are other aspects of it. The failure to find dark matter, and the rapidly growing evidence for MOND, is another. MOND works in galaxies, and it may or may not work in wide binaries (Chapter 130). At the moment, it's looking like MOND is an effect, rather than a universal gravity theory. But it's hugely important, and it still means the standard picture is wrong.

Neil Turok is someone in physics who speaks out, whether or not you agree with him. He's known for his frankness about the issues. He was for ten years director of the Perimeter Institute in Canada, in many ways the epicentre of theoretical physics. He describes the crisis in terms of our theories having got too complicated. He says our current theories fail to explain the *simplicity* of the universe. When trying to find any explanation, as William of Occam well knew, it's often a case of: simple is hard, complicated is easier.

In a speech in 2013, he described the crisis as being partly that some theories have not turned up in experiments. The Large Hadron Collider (LHC), strongly backed the standard model, but it left us with very little else, and the result doesn't narrow down the picture enough.

But worse, the LHC didn't find certain other things, including supersymmetry, for which confirmation was expected - and needed. Supersymmetry is a vital requirement for string theory. If supersymmetry is wrong, and it now looks unavoidable, string theory is wrong. The public generally don't know about this (it's returned to in Chapter 55), and no articles say that string theory was recently falsified by experiment. But although supersymmetry can never be disproved completely, the reality is, we've been shown that it's wrong. This hasn't helped with confidence either, to put it mildly. It now seems to many physicists that string theory and general relativity are both wrong.

Neil Turok puts the problem more positively than many who see it as a crisis. He paints the data as a set of clues - an opportunity for progress. But science thrives when we admit what we don't know, and he does:

The evidence is in. I think, by and large, all the wrong conclusions have been drawn by LHC and Planck. The physics world has continued on its merry way, although I'd say the discoveries made really seriously damage conventional viewpoints.

I find this really exciting because now it's time to put up another viewpoint. The hints that are there in the data are, I think, extremely bad for the multiverse picture. They're bad for super-symmetry, for string theory, and they're very good for the view that there may be a very simple, powerful theory that will explain all this. [...] everything turned out to be very simple, yet extremely puzzling - puzzling in its simplicity. We have to get people to try to find new principles that will explain the simplicity.

Some 'new principles that will explain the simplicity', are set out in Part 19. They're clear and straightforward, but they came drifting slowly out of a fog, through many years of struggle.

And for those who grapple with the difficult questions about time, it's not an easy path. Lee Smolin is one of the founders of the Perimeter Institute, and is respected worldwide. In his book 'Time reborn', he explains why he changed his mind, and decided time must be real and fundamental, after all. He's one of a number of us who are starting to see it that way.

One physicist, often found on discussion sites, recently called him 'confused'. But unlike many, Smolin has been out there on the front line, tackling the tough questions that must now be taken on. Others will avoid them, and feel more comfortable.

Paul Davies, who wrote a book about time in the mid '90s, said in the preface that after writing it he literally felt more confused than he had beforehand. It wasn't a turn of phrase - life on the front line is tough, and we should allow for that when people return from there without a complete conclusion. My own conclusion is certainly not complete in every way, but it's more so than most, and sets out a blueprint for how to move forward.

However well or badly we're doing, some of us are trying. And someone like Lee Smolin, whether he's right or wrong, should be given respect for that. At least he's had the courage to take it on. My own thinking on it started with confusion, and it took years to reach the clarity I later found.

Regardless of those who dismiss the problem, there's a growing realisation that these questions have to be taken on, and it looks possible that to solve them we may need the next paradigm, and perhaps even a revolution of the kind that Kuhn wrote about. If so, this new paradigm will probably arise, as special relativity did, and as other theories and paradigms have in the past, out of a response to a glaring contradiction.

43. A chink in the armour

So here's one - a response to a glaring contradiction. Although experiment is nowadays beginning to show our view of time was wrong, the problems with time still seem impossible to escape, and the rigorous logical reasoning that led to block time still seems unavoidable, and unassailable.

But if block time can *in principle* be wrong, that's a start. I realised during an exchange with Anthony Aguirre that people will need to know how it can be avoided. He seemed to think it impossible for block time to be wrong, due to what led to it. So I'll briefly show why in the new theory it doesn't apply, and come back to it later. The result solves two mysteries, and when you get to the second one, later in this chapter, it makes better sense.

And we've come to a turning point in the book. I've set out the puzzle now, and the problems with the present picture. I'll give a preview of my way out of block time, so people know a solution isn't ruled out before we start. And to be fair to those who believe in block time, which takes a lot of criticism in this book, I've got to say - it really does look like nothing could change it. So the attempts to squeeze one picture into another, although at times they do look unrealistic and worse, are more understandable.

What I call the three steps is a way to simplify the reasoning that led to block time. Roger Penrose, the well-known mathematician, came up with another way of simplifying it. The situation involves two people passing in the street, and an event in a distant galaxy. According to standard special relativity, if one person is moving away from the event, the other towards it, the event can be in the past for one, but in the future for the other.

This single point might seem interesting, and perhaps not too much else. *But it led to our entire view of time.*

A frame of reference (a 'frame'), is a viewpoint on how we take objects to be moving. In standard physics, frames don't do very much. But in what follows, they're slightly different from what we expect. It usually makes absolutely no difference, but in this case it does. There's a good reason why frames only work properly if you never take two of them to exist at the same time. And each frame relates two objects, not three or more.

The new theory is called PST, Planck scale time theory. As in Books I and II, in PST matter at its root consists of irregularities in the dimensions themselves: small waves in their fabric, which travel along the axes. But the axes have no fixed positions in space, they can be set at any orientation. So if you decide on a viewpoint for how matter is moving, you're implying a positioning for the dimensions, because matter travels along them.

But the axes can only be in one position at a time. So two different frames are like two different versions of the universe - they can't be taken together. If they are, you might get nonsense.

This hasn't been noticed because in special relativity two frames are hardly ever taken together. Instead one transforms (translates) between frames, as alternative viewpoints. That works fine: the axes being in different positions is never a problem at all *unless one assumes both frames exist at once.* But the reasoning that led to block time does exactly that.

According to PST, the scenario with two people passing in the street contains two incompatible frames. Instead of one frame with three objects, it is split into two frames, and each observer is related to the distant event separately. This means that the event is no longer past for one, future for the other. And that removes our present view of time.

The same point neatly solves another puzzle: the fact that light always moves at the same speed in relation to matter. This is a problem, but only if several objects are moving in different ways - in fact, it's what led to special relativity in the first place.

The problem is about relating the objects to a light beam, *and* to each other. We assume that the objects and the light must all be related together, in one frame. But in PST the light moves at c in each object's rest frame, each frame is incompatible with the others, so each object must be related to the light separately. If they're taken together, one is assuming the viewpoints can all exist at once.

This links the two puzzles. Both involve relating three objects. When relating a light beam to two objects made of matter, one way to avoid nonsense is to state as a principle, without explanation, the constancy of the speed of light in relation to matter: standard special relativity. It works well, but our system of frames really breaks down, as light can't be included. And yet we know a bit about photons - they're similar in some ways to other particles we relate via frames. So another way out of the problem is to say the system of frames was wrong, and that only two objects can be related at a time.

That means it's not a case of blaming Minkowski, Einstein or Lorentz, though by the 20th century the underlying problem came out more in a few areas. It goes back to Newton, Galileo, and intuition.

But if this is right, it's nobody's fault. There wasn't enough reason to think it might be that way. And to guess it would have been too far out of the box. I didn't - instead I got there via a roundabout route. It's explained properly in Chapter 107, and it shows how a single radical idea can suddenly make new sense out of two vital areas of physics, both of which have been unexplained, and have been seen as leading to inconsistencies, for a century.

Part 11. A hint from the simplest mathematics

44. The main clue in special relativity

Right at the heart of special relativity, you find a simple, unexplained setup. An object moving at a velocity v is seen with a slower time rate t. If you know its speed, it's straightforward to work out the time rate onboard the object, as seen by someone watching it go by. The equation for that is very simple - it's just:

$$t = \sqrt{1 - (v/c)^2} \ . \tag{1}$$

In standard units (SI units), v is the object's velocity in meters per second. c is the speed of light, also in meters per second. t is the factor by which the time rate is slowed, and is a number just below 1, as if the object's stationary time rate is 1. Time only slows down very slightly at lower speeds, so the equation gives a value for t just below 1, like 0.99999995.

So now here's the mystery. Of these three numbers labelled v, c, and t, two are speeds, v and c. The mystery is that the time rate t behaves exactly like a third speed, but no-one knows why. That was the mystery to me.

Several things suggest a speed. First, t is about the rate of matter's observed motion through time, which is already a speed of some sort, although it has a lot of question marks surrounding it. In this equation, it's the speed at which clocks are seen to mark time onboard the object when it moves, as a fraction of their speed when it's not moving.

And t behaves just like a speed, in particular in the way it varies in relation to the other two speeds. As I'll show, this makes a well known pattern, for some unknown reason. But t also behaves like a speed more generally, in the way that it corresponds to the *rate* at which change happens, as in the motion through time we observe in the everyday world.

It's hard to see how t could be a literal speed. But just to see what happens, let's try assuming it is one, and put it into units of speed like the other two speeds. Something interesting comes out. Many can see this at a glance, but I'll go through it and put all three into the same units.

In the equation, t is a fraction of 1. But treating it as a speed, the equation's form becomes even more familiar, it's simply $v^2 + t^2 = c^2$. This can be in any units of speed, with all three in the same units. To put t into the same units as v and c, the right hand side of the equation is multiplied by c, which gives

$$t = \sqrt{c^2 - v^2} \qquad\qquad (2)$$

you then get

$$t^2 = c^2 - v^2 . \qquad\qquad (3)$$

With everything in the same units, what comes out, $v^2 + t^2 = c^2$, looks exactly like the mathematics of speed components at right angles. It has the same familiar structure as Pythagoras' right angle triangle formula, but it's about the relationship between three speeds rather than three lengths.

On the face of it, this looks like something we know and understand well. A man running diagonally across a football pitch has two speeds at right angles, call them forwards and sideways. These combine to give the diagonal speed, so they're components of his overall speed.

Lets say his total diagonal speed stays constant, because he's running as fast as he can. If he changes angle, but then keeps running in a straight line, more speed will go forwards, so less will go sideways (or vice versa). If we call the speeds 'forwards', 'sideways' and 'diagonal', labelling them accordingly, then whatever angle he runs at, it's always going to be true that $f^2 + s^2 = d^2$. This is mathematically like Pythagoras' triangle, but it's about motion.

It seems that this simple picture is somehow similar to what space and time do, when matter moves through them. Special relativity looks very much like it deals with three speeds, and slips very easily into that form. The terms v and t behave exactly like two components of a fixed overall speed, the speed of light. And they behave as if they're moving at right angles - it's just that we don't know how to interpret that.

But thinking about it, there's one point that makes this 'speed components' version of special relativity far, far more likely to be the right approach than it would otherwise have been. It's the fact that the constant overall speed in question, c, is widely thought to be the general, constant, transmission speed for waves of any size, travelling through the fabric of space.

This means there could be some real physical setup underneath all this. The man running on the football pitch had a reason to keep his speed constant. In that instance, he was running as fast as he could. In the same sort of way, what we're looking at is a possible reason for the fixed overall speed in the equation: *it's the transmission speed of space*. The other parts of the picture are still missing, but perhaps not that one.

Anyway, whatever's going on, *c* doesn't change, but *v* and *t* vary so that the higher one goes, the lower the other goes. Knowing one, it's simple to work out the other. The general result is that the faster an object moves through space, the slower it moves through time, and vice versa.

This has been confirmed and measured very accurately, and it seems clear that it happens all over the universe. But why should space and time do that, or more to the point, why should matter's motion through them do that? It seems that the right angle implied in the mathematics means dimensions are involved. Right angles are one of the trademarks of the dimensions, and the dimensions are also very likely to be involved for other reasons. And it looks like some setup in which matter's motion through time and space combine as speed components do.

It's hard to see what's going on, partly because we don't know what motion through time *is*. But it looks like we might be seeing something at work here, some kind of mechanism, right at the heart of special relativity.

45. But *t* shouldn't be behaving like a speed

But *t* shouldn't be behaving like a speed. That doesn't look possible: a speed (or velocity) needs time in there, as part of how we define it. If *t* is a number that varies, and represents a speed, it's behaving exactly like the *v* in $v = d/t$. But if *t* has now become the *v*, what's that other *t* doing?

This may be a clue. The speed components picture is an analogy. Because the mathematics of time dilation is exactly identical to the mathematics of speed components at right angles, one avenue in trying to understand what's going on, is to say that what's going on is the same in both. Perhaps it is, but if so, the time rate's behaviour tells us a lot, and becomes very interesting. In the speed components picture, the time rate doesn't vary just any old how, like a varying time rate. It varies like a varying speed.

But if *t* does that, then it'll also be the *v* in $v = d/t$, and then you get another *t* coming in. A speed needs time underneath it, which means time would have to exist already. Surprisingly, this means that whatever the thing that we've been calling '*t*' is, it can't be time, *it has to be something else*.

Standing back a little, this could mean two things. Either it means the idea of *t* being a speed is a false avenue, along with the whole speed components interpretation. Alternatively, it means what we're looking at is a major clue. And what that major clue is telling us is that there's a deeper flow of time, underneath the varying one we've been looking at.

This path has a name. Well, it has one in this book - it's called the mechanism avenue. The mechanism avenue looks like this: you have an underlying flow of time, and some mechanism at work on top of it, which leads to the varying time rate t.

Two levels of time might look off-putting. But it's worth remembering that two levels of time are used in standard physics anyway (it's a different two levels - one frozen, the other with motion along the time axis). And one level plus a mechanism is far, far better than two contradictory levels. So if there actually is a mechanism to be found a bit further along this avenue, it would be worth finding, as it might solve the puzzle.

Part 12. Defining time, and a list of nine clues

46. Sticking to observables

It's well known that time is notoriously hard to define. That's part of why I have to include this chapter, and look at ways to define it. Existing definitions vary, but they often leave out two words that I'd absolutely need to define it - *matter*, and *change*.

One page on the internet (in the reference section) has a list of definitions, fifteen of them altogether. There are three from physicists and philosophers, and twelve from dictionaries and encyclopedias. Two on the list, attributed to the Oxford Dictionary and Google, are exactly the same - and that doesn't smell like a coincidence. Both of those sources define time as *"The indefinite continued progress of existence and events in the past, present, and future regarded as a whole"*. I then went 'control, F' to search the page, and found that, rather amazingly, both *matter* and *change* got 'no matches found'. The word *event* was used a lot, and came in eight of them.

My short form definition for time is *a continual, measurable, overall process of change to matter and radiation*. It's not complete, for instance you might want to add onto it 'that varies locally in a frame-dependent way'. But that's only for matter, not light, and you don't want to get dragged into explaining things like that in a definition - with definitions, you need to get in there, define it, and get out again.

It's for time in physics, and not like a dictionary, with 'time of day' and so on. 'Process' and 'change' potentially already imply time, but so did many of the definitions on that page. A lot of them use a pre-existing flow of time in the description, but we're defining *what we call time*. To me that's a hybrid thing with two layers: some points to suggest that follow.

The above definition is more than simply my own picture. Anything we don't understand is best defined by what's actually observed. And unless we can talk self-consistently about things we can't observe, such as spacetime and the four dimensions it contains (we've been having a bit of trouble with that, and we've now had to rethink our whole view of spacetime), it's best to stick to observables, and talk about what happens with light and matter. They go through changes: for instance, light's location changes over time.

So this works with or without the explanation for time I'll give later. The only definition on that page that has the wisdom (and well, gumption) to do the same, and stick to observables rather than including ideas, is Einstein's: *what clocks measure*.

Incidentally, this principle of sticking to observables can be applied in a wider way. There's a world out there, and then there are our ideas about it. Some people don't separate them, and it's arguable that you never can completely. But observables and ideas can still be very roughly, loosely separated. And in almost any area, separating them as far as possible, and looking at the world, at least briefly, without any presumed ideas at all - or trying to - can be very worthwhile.

Why do this? Well, in the past, our ideas have turned out to be wrong. Some think this was just up until the 20th century, but recent ideas are turning out to be wrong as well. In the 1970s, the standard view of life on Earth was that it was a freak accident, highly unlikely to happen elsewhere. This was deeply ingrained at the time - other ideas, like panspermia, were seen as heretical and crazy. But then, in just a few decades, we found thousands of planetary systems nearby, water in many places in our solar system, organic molecules in clouds of gas within our galaxy, and learned how hardy microbes are for travelling long distances perched on grains of dust, pushed gently outwards by radiation pressure from the sun.

So we now think that even if life first started on Earth, it's already travelling elsewhere. The idea that we're unique is just *gone* - we wouldn't be for long anyway. And life may drift through the galaxy, seeding any habitable planets. The NASA view is now that where there's water and the right conditions, life will probably appear. That's a huge turnaround, over a few decades. It shows that instead of presuming, we need an open mind.

47. The clues so far

Talking of sticking to observables, I promised a list of clues, and the emphasis in this list is very much on what we observe. Some of these clues will mean more to physicists, because they chime with other things we know. But their strength is in what they do when taken together - they're listed here because they belong together. And together they strongly suggest what we call 'time' is a hybrid phenomenon, with a fundamental element that doesn't vary, and a matter-related one that does.

So below are nine simple clues, all coming from well-known physics. They all point down the same avenue. Each clue has a chapter listed below.

Nine clues that point towards an unexpected conclusion: they suggest that time dilation is caused by a matter-related mechanism.

1.'Basic' special relativity tells us when an object starts to move, three things change in a mathematically identical way: its energy, its mass, its time rate. If the object, say, moves backwards and forwards, all three properties vary and shift in exact unison. Two of these are matter-related properties: its energy, its mass. So its time rate is behaving *exactly* like a matter-related property.

2. Confined light behaves just like matter. If light bounces between mirrors in a box, it gains three properties observed in matter, that it didn't have before, with the exact mathematics for each. First, its mass goes from zero to E/c^2, adding to that of the box, and mimicking matter's $E = mc^2$ pattern. Second, it takes on time dilation: the 'light clock' shows an object with internal standing waves is time dilated. Third, the mirror box when vertical responds to gravity as matter does: light zig-zags to the ground in the same time as matter takes to fall from that height. These similarities of matter to bouncing light suggest matter involves standing waves, at some scale where its nature is unknown. And the three properties light gains include *a matter-related mechanism that creates time dilation* - and it's an accepted and well understood one.

3. Time behaves differently in two well-tested and central theories. Quantum mechanics (QM) has time as an unchanging backdrop, while special relativity (SR) has time shifting around. So perhaps both have the same unchanging backdrop, but SR adds a matter-related mechanism for time dilation.

4. Point 3 is probed by asking if the SR rules are for matter, or universal rules. There's good evidence that they don't apply to light. Light breaks them, such as in having energy, but no mass. And it travels at c, with its energy small and finite. Matter cannot, but if it could, its energy would be infinite. So SR rules like relativistic energy and time dilation, which are mathematically identical, can only reliably be applied to matter, as *light breaks one of them,* ie. energy.

5. A well known aspect of SR is the 'speed components at right angles' aspect of it. Within that, the time rate behaves not only like a variable time rate, but like a variable *speed*. It behaves as one of two speeds that combine to make an overall speed, c. Whatever the cause of that, a speed requires time, which suggests a second level. So this looks like an underlying flow of time, with a varying matter-related mechanism on top of it.

6. Observations relating to time can only be made via matter. So matter may be unexpectedly central. Experiments show that as SR predicts, time dilation leaves permanent age differences between bits of matter. These cannot be explained via block time without *invoking unexplained behaviour of the time*

dimension. It was proved that if Minkowski spacetime is exactly right, there's no motion through time. But the age differences imply motion through time.

7. All attempts to 'generate' a varying flow of time via some mechanism, turn out to imply another flow of time underneath them - emergent time needs two levels. Deciding time is a dimension, on the face of it, is the only way out of this. But that turns out *also* to need two levels: one with motion through time, the other without. And they contradict each other (with the motion the randomness of QM demands an unfixed future, without it block time insists on a fixed one). Either way, two levels are needed. It turns out to be a choice between one level plus a mechanism, or two contradictory levels.

8. An object's total energy can be conserved, even if it has several different kinds of energy, and their quantities vary and shift within the total. Some see this total as purely a mathematical quantity, but if the total then changes, the object's *inertia* changes in proportion. So the total clearly seems to refer to something real. But if so, the different kinds of energy that combine to make the total are linked in some unknown way. This suggests fundamental, unknown matter-related physics at some level.

9. All matter also has its own basic internal energy, which behaves uniformly regardless of matter's state, and is unexplained. This also suggests across the board, unknown matter-related physics at some level.

Those clues are explained in these chapters:
1. *10, 17, 48*
2. *11*
3. *10*
4. *83 [para 5 on]*
5. *44, 45*
6. *9*
7. *32*
8. *8*
9. *91A*

Clue 10 isn't here, it's in Part 17, 'The best directional clue'. But far and away the most important *general* clue is the principle of least action, on page 237.

Some may want to go straight to the second half of the book from here, and the solution put forward there. For some, I may have shown the current view of time to be wrong enough for them not to need more. There's plenty that may be of interest before then, but if so, skim, skip or flip the next 30 pages, Parts 13 to 17, and get to Part 18, page 123.

Part 13. Putting time dilation into perspective

48. Affected or distorted: a paradigm choice

Before we get to some new ideas, there are one or two points that come out of the clues. There are also a few myths that need to be exploded.

There's one question that goes right to the heart of attempts to understand the universe. The answer is likely to affect the future direction of physics. It's about what happens to matter's properties in special relativity, when they shift around due to motion. Clue 1 puts it like this:

When an object starts to move, according to basic special relativity, three things change in a mathematically identical way: its energy, its mass, and its time rate.

Experiments show that matter's energy and time rate vary in just the same way. Mainstream physicists accept this. Many more also go by 'basic' special relativity, taking the view that mass also varies in that way, keeping in step with matter's energy via $E = mc^2$. This mass-energy relationship applies to matter everywhere else, so it would be risky to assume it falls apart.

The question I'm talking about is this. We know matter's properties vary, but when they do, are they being affected, or are they simply getting distorted? Some try to dismiss the effects of special relativity as a kind of distortion of what we measure.

This way of seeing it can involve the idea that there's some four-dimensional perspective type effect. The four-dimensional geometry of Minkowski can be taken in that kind of way. It has been a loose idea for some time, and has not been pinned down. One of its problems has been that it meant bringing in a second illusion to distort the first one, as one illusion simply wasn't enough. But some say that when an object moves, its energy, mass, and time rate are *observed* differently, which affects our measurements, and the mathematics of all three distortions to the results is the same.

On the face of it, this does look possible. The mathematics of these changes is loosely similar to that of perspective. In fact it's different in some revealing ways, and that particular bit of mathematics fits many phenomena. But we can say this perspective idea might or might not be valid. So we now urgently need to distinguish between two scenarios.

This is potentially a question, without wanting to overdo it, of paradigm-shift level importance. Looking closely at this 'affected or distorted' question may show something crucial about how to understand the universe.

Let's start by thinking about how a visual picture can get distorted. If a pencil is held horizontally, and then angled away from someone watching it, it looks shorter. But if you measure its length in three dimensions, it'll still be exactly the same. The way we see the pencil is effectively as if it's in two dimensions, like in a photo. So angle it away from a camera, take a photo, then measure it on the photo, it really will come out shorter. The measurement has been thrown out by a perspective effect.

Perhaps something similar happens when our object starts to move, and its energy, mass and time rate are all changed. If so, a perspective effect would be going from four dimensions to three, instead of from three to two. But it doesn't fit that time goes slower, while energy increases. That doesn't look like they're being distorted together. Remember gravitational time dilation, the other kind, goes the other way - time goes slower, and energy decreases. That *does* look like they're changing together. So this weakens the distortion idea.

But that's not the main point. Returning to the pencil, rotate it back through the same angle to its original position, and the effect disappears completely. That's how we know there was nothing more than what we call a perspective effect. So one way to distinguish between 'distorted' and 'altered', is to see if the effect leaves traces behind it. If it does, one of the pictures is wrong. And we know one of these effects leaves traces behind.

Time dilation makes indelible alterations to the universe. Two particles might move differently for a few seconds, and they can then swim alongside each other for billions of years. The age differences will remain, and will become a built in feature of the universe, visible from any viewpoint.

So these changes are real, not distortions. All three effects are similar: time rate, mass and energy all go by the same mathematics, and vary together. So if one of them is real, we certainly can't assume the others are unreal. So the idea that these effects are distortions is holed below the waterline.

49. Why do things get 'angled into the time axis'?

The everyday 3D perspective effect is well understood. When the pencil is held horizontally, then turned away from the viewer, it looks shorter. Turn it back and the effect vanishes. The temporary distortion affects distances, and we know why. Parts of the pencil that are further away look smaller, because of what happens with the light that travels to the eye.

The 4D version of this, if it is one, is very different. It alters the pencil's time rate, not its length. But this effect isn't set off by turning the pencil through an angle. For the 4D effect, you throw it across the room. For some reason, the pencil would then be seen 'angled into the time axis', as it flies through the air.

This is unexplained in the perspective analogy. It was one of many things that just had to be accepted as part of the nature of the time dimension. In 3D perspective we actually physically angle the pencil in the direction of an extra axis, to get the effect. But in the 4D version, if it is that, we have to assume that something else, which we have no understanding of, somehow causes it to be angled into an extra axis, *when it moves*.

As it flies through the air, the pencil 'moves through time' slower than other objects in the room. We know this happens - the effect has been measured. Perhaps it's because the time axis now 'looks longer', in roughly the way that in 3D perspective, one space axis looks shorter.

But other things happen as well. As the pencil flies across the room, its mass and energy increase briefly, flaring up and dying down again. When it lands on the carpet, they quickly go back to where they were before it was thrown. Something is 'going on', that much seems clear. When the pencil lands, its atoms are older than before, but the other atoms in the room are more so. They've been ageing faster than those in the pencil while it was in flight, and the age differences will be there for as long as the particles exist.

We don't know why this happens. But even if we were to accept that there's some weird kind of perspective effect, it *still* makes the time axis look like it has motion on it 'built in' somehow, as what gets distorted is time speeds, not distances. The whole setup behaves as if the fourth dimension is not time, but motion through time. So if the motion is emergent, as some think, we shouldn't be finding it hooked in with things like space dimensions, Hermann Minkowski, and 4D geometry.

The perspective interpretation is a way of trying to look at time dilation as an illusion, but it doesn't include the actual timeline motion, which slips through the net, and comes out looking real. The fact that this motion gets distorted (in one way or another) has allowed us to zoom in and take a look. And the differences that remain afterwards look like the result of a real flow of time being distorted.

And the perspective idea falls apart: normal 3D perspective is a visual effect that takes something *real* (distance), and distorts it temporarily, leaving no traces. But time dilation, whether or not it's a 4D version, takes something thought to be *unreal* (motion through time) and distorts it. And yet it does leave traces afterwards.

So the perspective analogy only has a chance of working if one part of what it might explain, the timeline motion, escapes - from being explained in that way.

And if there are illusions at work, you need two: a perspective type one that we have not much understanding of, and another for motion through time that we don't understand at all. And if these two illusions interact in exactly the right way - a physical illusion distorting a psychological one - perhaps you get what we observe.

But the trouble is, if you do that, the interaction of the two illusions means the other 'illusion', for motion through time, then pops out of where you put it before, and starts to look real again. This whole set of clues seems rather resistant to being dismissed as illusions.

Part 14. Speeds or distances?

50. Living in a field

Like the question in Chapter 48, 'affected or distorted', here's another one. 'Speeds or distances' is about which gets altered on the time axis. Like the last question, it affects what the next paradigm might look like, and there are pointers about it in a few places.

The question is about what varies when time dilation happens: time speeds, or time distances. If it's time speeds, motion through time looks real. If it's distances, that looks like block time, in which the time axis is static. And as I said on time dilation, 'the whole setup behaves as if the fourth dimension is not time, but motion through time'.

Many physicists know this, although the mathematics can be trimmed down until some start to hope the differences will go away. If they did, we'd be left with distances, just like on a space axis. But the truth is, the time dimension behaves as if timeline *motion* is what the dimension is. But no-one says that, because the motion isn't meant to exist, and there's nothing that can be said alongside it.

I'll list the places where this happens: the first is time dilation due to motion through space, which was looked at in the last chapter. If you try to put in a perspective type effect, the timeline motion survives the process, and comes out looking real.

And second, the other kind of time dilation does the same. A quick look, and gravitational time dilation can be seen distorting time speeds, not distances. Again, this makes the speeds, and the motion, look real.

In the well-known 3D analogy, which helps with picturing general relativity, a bowling ball sits on a trampoline. It represents a mass such as a planet, and the sheet gets curved off into an extra direction, downwards. Each point on the sheet changes its *position* by a certain distance, along the trampoline's up-down axis, which is an extra axis at a right angle.

If an ant walks along the sheet, the nearer the bowling ball and the centre of the indentation he or she goes, the further that particular place on the sheet has been moved downwards.

But what happens in the actual 4D picture from general relativity, with space curvature? Instead of a 2D rubber plane curving off into a third direction, you have a 3D space curving off into a fourth. Both the plane and the space will normally be 'flat', unless a mass distorts them out of shape.

And in general relativity, the fourth direction - off into which the curve goes - is time. Without going into the details (the picture is not complete in general relativity) both space and time get distorted out of shape near a mass, and in similar ways. This holds some real clues. Two different *kinds* of dimension get distorted in the same way, so we can take a peek.

And the time axis is the equivalent of the up-down axis of the trampoline - it behaves like a fourth large-scale axis. So although this is only a loose analogy, it gives us a framework, and it's interesting to see what happens.

And what happens isn't what you'd expect. It isn't like what happened to the ant, when it was shifted along the up-down axis. Time dilation doesn't shift points in space along the past-future axis, changing their *positions* in time by a distance, a little way into the past or the future - as one would expect if the trampoline analogy was a close one.

What actually happens is that the *speed* at which time passes is altered, and in a way that's related to the distance from the centre. The nearer the mass you go, the slower time is passing. It's not distances, it's speeds, and the two are often different, conceptually and mathematically.

So what the ant experienced on the trampoline (it's not anti-gravity), doesn't happen. The ant found distances had moved, but instead it's speeds. So here we have a useful area. We can see the same distortion affecting both space and time, so we can compare them. The mathematics of the distortion to the two kinds of dimension is similar, but what actually happens tells you about what gets distorted. This comparison between *kinds* of dimension is packed full of clues, and if you look closely, you find the timeline motion is what gets distorted. So we can't add it in later, as an extra.

51. Elapsed time experiments

The gravity kind of time dilation, like the motion kind, leaves permanent age differences behind it. Lasting traces remain after an elapsed time experiment in which two clocks, or two people, spend a few hours at different heights in a gravity field. Their ages come out just as if time speeds have been different. This was measured to a new level of accuracy at NIST in 2021.

Say one astronaut stays in orbit, and another goes down to the surface of a planet or moon. This has actually happened, with the moon landings in the '60s and '70s. Two would go down to the surface, one would stay in orbit. Comparing one astronaut from the surface and another from orbit, there'll be a new age difference between them, just as if time speeds were different at different heights in the moon's field.

A believer in block time might say that only the length of time *intervals* (time distances) are different at different heights, and not time speeds. I've known good physicists to think that, and to see the speeds aspect as some illusion. In the view that only time intervals are different, it's thought that both of the astronauts exist at many moments throughout their history, alongside each other. The distances across these points are larger for one astronaut than for the other, so in a sense the two lines get out of step with each other, but not because of anything moving along them.

Underneath this is the idea, from general relativity, that not only space gets curved near a mass - the time axis does as well. This curving of the axes is thought to be real, not an illusion.

On the face of it, this seems possible - one way to see it is that near the mass the time axis is stretched out longer, into a curve, so anything takes longer to travel it, producing the set of time rates that we predict and measure. So one astronaut gets stronger 'timeline curving' than the other, and he was nearer the centre of the moon's gravity field.

So far the picture works, and in places it makes perfectly good sense. You can make the time *intervals* the equivalent of space distances, and the two kinds of dimension then look similar.

But now try adding an illusion into the mix, to explain the timeline motion we actually observe. This part of the picture has not been thought through so well - a psychological illusion isn't our field. But the illusion, for those who believe in it, is added onto the gravity field as an extra, and only where there are living creatures to observe things.

It then starts to get very far-fetched, given that it's part of the standard view. The gravity field, without living creatures hanging around, has nothing but a pattern of *distances* along the time axis. As with the space distances, these are graded, so the nearer the mass, the more the time axis has been bent out of shape. Both space and time distances are stretched, as if everything has to take the long route across the curved floor of the indentation, rather than travelling in a straight line.

But now we get to the unconvincing bit. According to the illusion idea, when it is observed by living creatures, this pattern of distances looks like a pattern of speeds. The pattern of distances gets fed into the illusion, and the illusion then obligingly converts distances into speeds. You get a graded pattern of time *rates*, because living creatures see an illusion which makes everything move along the time axis.

This explanation is so bad, it's embarrassing. A gravity field is like a synthesis of several closely linked effects - but if we're to believe there's an illusion at work, we have to believe one of these effects is totally unconnected with the others, and with the gravity field itself. Instead of being part of the field, it just happens to land on top of the other effects, wherever living creatures evolve. They then see this weird set of adjustments to their illusion, and they later evolve to the point where they can measure them.

We're good at skepticism in the 21st century, and it's time to get it out and dust it off - a bit of healthy skepticism is appropriate here. It's far more likely that the complicated pattern of time speeds that we measure near a mass is directly connected to the gravity field, and a part of it somehow.

Anyway, if there's an illusion, it would certainly need a physical basis, as well as a psychological one. With a shared illusion that we all agree on, such as a rainbow, there's going to be something physical going on. Without that, it's hard to see how it could be so precise. In the Earth's field, near the surface, the time rate differences mean that the length of one second varies by about 10^{-13} seconds for each kilometer of height.

The illusion approach looks even worse when one finds it isn't supported by the other kind of time dilation. And that's one of the very few areas where it might find some support.

Time dilation due to motion also has the illusion coupling onto other effects in an unconnected and contrived way - but in a different way. When people add an illusion into the picture, neither kind of time dilation gets any support from the other. *They arrive at the same result in different ways.*

This century physicists gave up trying to convince themselves of the illusion idea, even though we spent the 20th century trying to believe it, and easily convincing the public.

Emergent time, which replaced it this century, has similar problems. Where time varies its speed, it's hard to insert it into both forms of time dilation in a way that is self-consistent everywhere. And with no psychological element, emergent time can't explain the apparently perception-based aspect of time, which came from Minkowski's spacetime picture, and is what set off all these desperate ideas in the first place.

That seemingly perception-based aspect of things is where illusion time does a bit better than emergent time - or it would if it worked. But both fall down, along with everything else, in a gravitational field.

52. The grain in the block

Block time is officially a part of the standard view, so it's fair game, and very reasonable to show where it fails. But I'll move on, hopefully having shown it up for what it is. The truth is, people assume that things we don't understand are at work, and think 'it must be something like that'. But some still talk to the public as if we understand time.

When we try to imagine time dilation in the context of block time, all these time distance variations would have to be frozen in the block as well. Those two astronauts in the moon's gravity field would be set in the time slices in the block, stretched out, with their ages getting further and further out of step.

So this four-dimensional block, if it existed, would have built into it a strange uneven 'grain', because bits of matter are ageing at different rates. And both kinds of time dilation do it, leaving patterns in the block.

And this uneven grain, if one stands back, suggests not only motion, but laws of motion. But if nothing actually moves through the block, as was 'proved' in the 1960s, it's then hard to explain why this grain is there. As always, people assume it's explained by two separate levels, but I hope I've shown this can't be so.

So with all these unconvincing aspects, why is block time part of the standard view? Because block time is an unavoidable consequence of spacetime, and many hope spacetime is right. They don't want it to be wrong, or we have to admit we have no standard view.

Block time is seen as uncertain, and opinions differ. It has been questioned recently, now spacetime has started to look shaky in several ways. But some areas of block time are like rooms in a house no-one ever goes into. I think I've gone into them enough to show you why.

When looked at closely, all of this doesn't look like a frozen, static universe. Quite the opposite, when all the clues are taken together, all these time rate variations make the universe look alive and dynamic. And that fits with the dynamic picture from quantum theory, in which time is not frozen, and the future is wide open, because of fundamentally random behaviour. It leads to a world that's more exciting than the frozen one, as it's full of unforeseeable, unpredictable situations.

53. Which direction might have a concealed way forward?

So both kinds of time dilation show speeds, not distances, being altered. And once again, we hear the rumble of unexplained motion on the time axis. This 'built in' motion turns up everywhere. And it's not false intuition, though this has often kept people away from looking closely.

This motion is hooked in with the geometry of dimensions. And that brings me to the speed components picture, which is another area. And once again, if one looks closely, one finds speeds, not distances.

As things are at present, we can take the timeline motion to be real, or take it to be unreal. And turning one way or the other, neither direction is an easy road. So the question is: which direction is more likely to have a concealed way forward, that is, something we can't yet see.

And if you compare them in that way, it's no contest. The superficial time approach is a set of 'desperate remedies' - a fix-up measure invented partly to rescue a flawed picture. It exists to protect special relativity from the fact that, although the mathematics is right, our understanding of it looks wrong somewhere. But most physicists think if we're to fuse the two main theories, something will have to give *somewhere* anyway.

And it's likely to be in the interpretation department, so a flaw shouldn't be all that surprising. Meanwhile, in terms of the picture, the standard view has reached a brick a wall, with nowhere to go. But there's also the idea that this motion is somehow real, and exists at a deeper level than the physical laws that depend on it.

And looking in this other direction, instead of a brick wall, you see there's a rough path, a little overgrown - but which might perhaps go somewhere. And it's clear that this other way might have possibilities.

Part 15. Planck scale physics

54. String theory

In string theory, all matter is made up of very small objects, *way* smaller than the particle scale. It's a scale far, far away, but it's also right here. In the mid-'80s, people started working on what happens there.

A closed string looks like a rubber band, it's a circular loop of matter. They're all the same. They vibrate in different ways, to make each type of particle - it showed that all matter might have a common root. 'The Planck scale' is often used to mean a loosely defined scale beyond the particle scale, perhaps at or near the literal Planck scale - as it means in the theory in this book.

Although its basic ideas may be loosely true, this century string theory looks disappointing to many, though it seemed very promising near the end of the last. It used to look like a path towards unifying all physics into a theory of everything - people thought there'd be a way through. But thirty years on it has problems, and is incomplete, including conceptually. Until the mid '90s it was a group of five theories. It then split into a huge number of them, with no way to pick the right one. Many are critical of string theory, but despite its problems, it has shown what a way forward might look like, even though it's having trouble providing one.

It has shown a way to simplify our understanding of matter. The objects that create everything live at the same scale where there are thought to be curled up dimensions. It's a strange place, and our picture of what goes on there is still forming. We can only reach into that far hidden world with two things - mathematics, and the human imagination.

Nevertheless, the world we know somehow seems to arise out of whatever exists at the Planck scale. At larger scales, we find different units of matter in different relationships, all the way up to superclusters and walls of galaxies. But it all somehow starts from the same root, at the Planck scale. So perhaps string theory isn't getting there because we lack some concepts - it's likely, and people say so. If there were physics courses that focus on the conceptual side, progress might be very much quicker.

Looking at how physics developed, there's a pattern. During the 19th century, chemistry got increasingly complicated. Many new substances appeared, the periodic table that listed them grew - we mapped out a complex landscape, filling in all the details. But meanwhile, a far simpler world at a smaller scale was starting to appear underneath it: particle physics.

Then in the 20th century, that simpler world itself got more complicated, and a whole new landscape emerged, with a growing zoo of different particles, in all its diversity. Again we filled in the details, mapping out a new level - the new periodic table was a list of particles: the standard model.

Then the same thing happened again. By the 1980s we started to find hints of a simpler world at a very much smaller scale, that was beginning to appear underneath everything we had: Planck scale physics.

So as we zoom steadily in, new layers keep appearing. We seem to be going at a rate of about one layer per century at present, but the Planck scale looks like a very fundamental boundary. There may be no new layer for our great great grandchildren to work on in the 22nd century - but I'm sure there'll be plenty for them to work on anyway.

55. Major theories are an endangered species

Part of what has happened in physics this century is that our excellent new measuring devices are showing our old theories are wrong. Special relativity and quantum theory are fine - everything else is in trouble. There are a few facts that many don't know about physics in the early 21st century. The first is that general relativity is now seen by many as ultimately wrong. The second is that string theory has turned out to be wrong.

String theory needed supersymmetry, which the LHC didn't find, to save it. These large theories are failing in the light of new, accurate data, although nothing 100% certain has appeared. But it means the present paradigm has shifted, from solid ground to very shaky ground.

And Roger Penrose, who is highly respected, has said that inflation theory has also been falsified. Inflation has major problems (Chapter 126). But it's fiercely defended. Why? Major theories have smaller theories to save them. Just as string theory needed supersymmetry, GR needs inflation.

We know space is flat at a large scale, from the CMB. A lot of theories expect that, so for them it's no problem. But GR has to *explain* why there's no large-scale curvature, because it predicts curvature. So it needs the universe to be at the critical density. We're told it is, but the main reason for thinking it is, is assuming GR is right. So there has been a bias towards things that help get it near the critical density. These are not fix-ups - people just think it must be so, as they think GR is right. But even with the add-ons to take it nearer, only inflation can make $\Omega \approx 1$ a real possibility. Otherwise, with all that expansion, any miniscule divergence from that density would be amplified: by now the universe would be nowhere near the balance point.

But with inflation, the rapid expansion pushes space out so much, it ends up flat. This idea is needed to save GR. So it gets protected: Stacy McGaugh talks about how lambda was brought back in the '90s, partly to prop up inflation. But now Penrose says inflation has been disproved.

Let's return to string theory. Supersymmetry is not a theory, it's a large class of theories: technically it can never be ruled out completely. It's needed now to prop up the present paradigm in several ways, including providing a range of new particles, which people hope will explain a serious problem about the Higgs mass. It's also needed to save string theory.

Some say string theory can't be tested, but in fact it already has been, and it failed. The LHC was very much expected to find at least some of the particles of supersymmetry, and people didn't worry about it. But this was actually a prediction that string theory quietly depended on. By 2013 it was clear that supersymmetry isn't there - almost certainly. But if it isn't, string theory is wrong. Since then, many have stopped working on it.

But we can still look at the picture. It might be half right, and it's full of clues. String theory unifies all matter, but the building blocks are still thought to be made of 'stuff' - that is, matter. In theories of matter, people try, but so far it hasn't been easy to describe particles via the local behaviour of space. But many, including Einstein, have said we'll eventually be able to do that, and reduce everything to space: the knots and hills of spacetime. But for now our theories describe matter as separate from space.

This leaves open the question of what matter actually is. In the last century, people wondered if the pattern of particles dividing into smaller and smaller units goes on for ever, or if the sequence stops somewhere.

The Planck scale, which we're approaching now, is where our ingredients for the universe appear. What we call space and time begin there. So for many, it's an exciting time. It's interesting that matter seems to arise at that scale: if there's an origin point for matter sitting there, we might be near to finding out what matter actually is. So reading on, you might find out what I'm made of - that's one way of putting it.

Part 16. Mythbusting

56. The arrow of time

A friend of mine once wrote a song called 'Questions, no answers' which was serious and light-hearted at the same time. He played it at a Zoom acoustic night. The questions he listed in the song included 'Why doesn't time go in reverse?' The song was fine, and it didn't matter, but the question is actually a misleading one, or rather, the emphasis on it is.

That question was more relevant before experiment showed the single idea that led to it was wrong in the first place: the idea that small-scale physics is reversible. But even without that, one of the most common false ideas given to the public is that the laws of physics are time-reversible.

In the equations describing small-scale events, some things look potentially time-reversible. It's not everything, and it's certainly not a rule. But down at the particle scale, a lot of mathematics seems reversible. On the other hand, vital processes down there are irreversible. That includes the 'collapse of the wave function' in quantum mechanics - the change from waves to particles. It's often left out when discussing this, but it's a major part of the universe, and it's utterly irreversible - except via massive coincidence.

At larger scales, in the real world, cause and effect makes things irreversible anyway. We now know this is true at small scales as well. So this point about reversibility is just about the mathematics.

The mathematics is an incomplete version of what it represents. It's a bit like an mp3 file, or a compressed computer file, that has some of the information removed. But some who give importance to the mathematics beyond the call of duty - or of reality - see the reversibility as significant, and have asked why looking at the real world, time only goes in one direction.

The arrow of time means a pointing arrow, that indicates a direction, rather than a moving one. It's about the consistent direction of the flow of time we seem to find around us. This 'one way street' aspect of time is unexplained, and some who try to uphold block time imply that it's the only question left to answer.

But it's a smallish mystery sitting next to a massive one: the apparent motion through time, or flow of time. There are some who ignore the large mystery, and focus on the small one. At times it's a way of trying to diminish the real mystery, and draw attention away from it. But we'll probably only know the answer to the small mystery - time's direction - when we solve the large one - time's apparent motion.

Some people try to explain the single direction of time by attaching it to one of the irreversible processes we know, that create an arrow. Something that makes an arrow of time is anything that develops in one direction. To some who think there's no flow of time, it seems that some other arrow of time must come flying in and take the role. But as we'll see, nowadays this doesn't fit the facts.

Attempts have been made to link the arrow of time with a process we know goes in one direction only: thermodynamics, or the increase in entropy. No direct links with it are known, but it has one thing going for it - we can't think of anything else.

The second law of thermodynamics deals with how heat exchanges between bits of matter tend to smooth away patterns or structure over time, reducing the amount of order. Thermodynamics goes in one direction on average, but it goes backwards and forwards everywhere locally. The public get the idea that thermodynamics explains the *flow* of time, but it's used to try to explain the arrow, not the flow. It only provides an arrow on average, it provides no flow of time - and it needs a flow of time underneath it to work.

Physicists tend not to highlight this point, but thermodynamics generally only becomes relevant when one has already decided that motion through time doesn't exist. If one decides that, the next question is what causes the arrow - the *direction* of time. If we knew we could see all there is in physics, with all the jigsaw pieces in front of us, we might wonder if thermodynamics causes the arrow. But we're nowhere near able to assume that.

Apart from thermodynamics, there are plenty of other processes that only go in one direction, although many of them could in principle go backwards. But that doesn't mean much, as in fact they never do. An important one is wave emission. It happens at all scales, all the time: the waves go outwards. Throw a stone into a pond - the ripples will travel outwards in ever-widening circles. In principle, via the conservation of energy, all the ripples might reconverge right where the stone will soon emerge from the water. The energy released earlier could fling it out of the water. But no-one sits by a pond waiting for this to happen, because it probably won't happen even once in the history of the universe.

Some talk as if this hypothetical reversibility - in principle - of wave emission, is a point in favour of that view. But if very long odds are allowed, you can do anything you want. Massive coincidence is a highly useful tool for reversing things - it *really* helps. We could go around the universe fixing situations. But back in the real world, physics goes one way only.

And we also find irreversible processes at the particle scale. When Eddington linked the arrow of time with entropy in 1927, quantum theory had only just arrived, and the basic irreversibility it brings had probably not been digested. Then there are other irreversible small-scale processes, which added in make thermodynamics look even less significant. Among them are kaons: particles that break the reversibility in the way they jump around.

So if the flip book of the block universe is flipped backwards, looking at large-scale - and small-scale - laws, you don't get much that resembles sense. You also don't get much that resembles physics, as it actually is. But what you *do* get more often, is something that resembles the mathematics.

But mathematics is a shorthand version of the world, a lot of information is removed - and in this case, enough to give a false picture. And underneath all this is one very central fact: time, in the mathematics, is like space. But in the picture, time isn't like space. So people have tried to 'spatialise' time, and to squeeze the conceptual side into a configuration it can't possibly go into, to accommodate the mathematics. It's like a foot crammed into a shoe that it simply doesn't fit into.

But the mathematics of time is simple: it's speed components at right angles, like Pythagoras' triangle. It's a pattern that gets hidden in teaching systems, and people are often told that it's far more complicated than it is. But that mathematical pattern appears in other places in the real world - there's a list of things it describes. So the unexplained mathematics we've found could be describing a whole range of possible pictures.

57. Thermodynamics and time

Ethan Siegel, who I sometimes agree with, points out common errors well (as in Chapter 125). In an accurate article *'No, thermodynamics does not explain our perceived arrow of time'*, he rightly says there's no link at all.

It's true that thermodynamics is different from a lot of processes. The second law applies widely, but it gets reversed locally. Temporary order is created, but then compensated for by disorder appearing nearby. Crystals form, order increases locally, but outside that area, you get extra heat.

Only looking at the wider picture is entropy irreversible, *on average.* If you look locally, there are pockets of order in many places. And in them, time's arrow won't match up with entropy. Time is precise, local, and it always goes in the same direction, not just on average.

But the whole area is vague: some think the link between thermodynamics and time works locally, others think it doesn't. An American professor who I once asked about the entropy idea said he'd heard as many views of time as he had friends with whom he'd discussed it. There are also other areas that are questionable relating to entropy, such as issues about the early universe, first pointed out by Roger Penrose.

Thermodynamics can be pinned to the direction of time, but only in one way. The list of things it leaves unexplained resembles the entire list of questions about time. But it does do what Eddington was doing when he first started the idea. He was interested not in the *cause* of time's arrow, but in a way of officially telling which way it points: which direction goes towards the past, and which goes towards the future. Entropy certainly does offer that, it's a good indicator of it. But it offers nothing else. With its meaning changed, it becomes a broom to sweep the time issues under the carpet.

It doesn't explain motion through time, as some articles imply. Physicists, to be fair, don't say that, they only say that it creates the direction of time. But it can't even do that. In the true block time picture, thermodynamics doesn't exist. Heat depends on particles moving around, but if they're to move *at all*, they need time already in place. If nothing moved through time (as has been 'proved'), the universe would be at absolute zero.

Realising you can't have what many quietly believe - two separate levels - it's oddly simple. No motion through time, no motion through space. No motion through space, no heat. No heat, no laws of thermodynamics. Make motion through time superficial and emergent, you have to make thermodynamics more so. It can't be *creating* a direction of time. Without some flow of time, with a definite direction, it wouldn't exist at all. So it's an effect, not a cause. But if you have two levels built into the time dimension (constantly implied but rarely said), the problem seems to go away.

But another group of physicists think differently. To us, there's no need to try to explain this 'one way street' aspect of time. And any link between entropy and the arrow of time - other than Eddington's original one - is unconvincing in other ways.

It looks unconvincing because of the timeline motion that's implied in many places in physics. It behaves as if it runs deep: when the underlying geometry shifts around, it does too, so it looks closely associated with the dimensions.

So rather than dismissing this motion or trying to explain it away, some think we need to understand it. The consistent direction is just one aspect of it - what we need to do is to explain the whole thing.

58. An experiment on the arrow of time

A new and potentially fatal flaw has appeared recently in the idea that the thermodynamic arrow explains the so called time asymmetry. That whole set of ideas started from a single idea - the apparent reversibility of physics at a small scale. That's what set off this whole approach.

But we find it isn't true. In results from a series of experiments done in 2015, it was shown that there's an unexplained irreversibility even in the quantum world, right where the physics is meant to go in either direction.

The whole idea was that at a small scale, you just have reversible processes, and that's the real, fundamental, underlying physics. But at a larger scale and at a more superficial level, the direction of time *emerges* due to increasing entropy. But now we find out that the directional entropy goes right down to where the reversible physics is meant to be. So we can't just say it emerges further up the scale. And it seems that those reversible laws, to which people attached so much importance, are an incomplete description of the world. It seems the small-scale world isn't reversible after all.

This result is still being digested, and it's early days, but it may be crucial. The experimenters had some carbon-13 atoms in liquid chloroform, and flipped their nuclear spins using a magnetic field. They then flipped them back again, and if the world is reversible, they should have gone back to how they were. But they didn't, and repeating it rapidly many times, the whole process was shown beyond doubt to be affected by entropy.

In their discussion at the end of the paper, the experimenters ask what they call the 'puzzling question' of how the Schrödinger equation can lead to a strictly non-negative relative entropy. And one of them, Mauro Paternostro of Queen's University, Belfast, said in a magazine interview: "*Our experiment shows the irreversible nature of quantum dynamics, but does not pinpoint, experimentally, what causes it at the microscopic level - what determines the onset of the arrow of time.*"

As always, it takes a while before the implications filter through, but it leaves that whole explanation, with time asymmetry arising from thermodynamics, looking even more like an inadequate fix than it did before.

That's because something, somehow, is causing time down at that scale. And it's looking recognisably like time as we know it in our large-scale world. This suggests that time is caused by something we simply don't understand. And whatever it is, as far as we know it's causing time at all scales. So it highlights the need for a new view of time. I'll return to the experiment in Chapter 117, when I've suggested one.

Thermodynamics has been about larger scales so far, and it hasn't been fully integrated with quantum theory. But people are working on it. It's not only experimenters, recent theoretical work suggests that the second law, rather than falling apart, holds well at that scale.

That experiment is potentially another first domino. It leads to more ideas falling, and eventually wide areas of the standard view. It's the equivalent of the near-proof in Book II (and Appendix D here, *'The proof of the pudding'*), which shows beyond reasonable doubt that gravity works in a different way. That little bit of mathematics can be set out on one page.

But like the little experiment above, it's enough to bring down the present paradigm. It's significant for its small size, but it can't be compared to David's slingshot, or what they managed to fire into the exhaust port near the end of Star Wars '77. Instead it's just a pointer to the truth, part a general search for truth - but a very clear one, and very compressed.

The present view of time is failing - including in the lab. I'd say it's a view on the way out, but that may depend on what happens. Until it gets replaced by something that works far better, it will stay where it is, doing what it does - stopping progress by blocking a gaping hole in the jigsaw, convincing enough people for it to keep sitting there, and making it less likely that we find a true way to fill the hole, and a real way forward.

59. One second per second?

Before showing you the very best clue, I'm still showing flaws in the present view of time. The public find it easier to believe the present view than many physicists nowadays. To some who don't know the real landscape, it's easy to dismiss motion through time. Physicists don't do that any more: we've been grappling with the issues and problems closely recently.

But philosophers sometimes do, and may get the impression that a flow of time can be disproved. One critic of the idea said: *'The idea that time can in some meaningful sense be said to flow, is just a complete non-starter. For time to flow, it must do so at some speed. But speed is measured as a change over time. So how fast does time flow?'*

The second sentence in there is nonsense: *'For time to flow, it must do so at some speed.'* That's inappropriate mathematically, because speed contains time - it's false. So before we know how time works, he immediately applies the wrong mathematics to it, and when that doesn't work, assumes we can draw conclusions from the fact that it doesn't.

In fact we do know how to apply our mathematics to time. It doesn't involve assuming any absolute speed, as he implies it would. We routinely measure flows of time *in relation to other flows of time*. That's special relativity as it actually is, rather than in our ideas about it. That's what the equations do. We also do this with time rates in general relativity, and it's an observational fact, having been confirmed by experiment. Only in our ideas do we remove the flow of time we observe.

And if you look at physics as it actually is, we're constantly measuring things in relation to other things. The motion of objects through space has been known to be relative for centuries - a lot of things can only be measured that way. And with time, not only *can* one time rate be measured in relation to another, it's an absolutely fundamental part of our physics.

If there *was* a flow of time that existed in some absolute way, then however it gets described from the point of view of observers like us, straddling bits of matter within the flow (and whatever matter is in relation to it), it certainly couldn't be described in the way he mentions - that far we agree.

Giving time a speed is a kind of circular reasoning, so it shouldn't be brought into the discussion. It's like a straw man to be knocked down easily, and it stops people thinking about the real questions.

Above all, and here's the vital point: it can be made to seem as if a circular argument comes *out of* it, such as the idea that time flows at one second per second. But in fact the circular reasoning was put *into* it at the start, so of course you get nonsense coming out. That's what's underneath this point, if you look at it closely.

Anyway, giving time an absolute speed, as the person who said it might well agree, is an invalid approach (if the word 'time' is used as it's normally used in physics). But he'd have trouble arguing that it's the only possible approach to the question - there are others, and they're far better, as in the way time rates are compared in special relativity. So in saying it's the only approach, as he does, he's simply wrong.

I hope this section has exploded a few common myths. Before moving on I should mention cause and effect. The reason there are so many false ideas and myths around time is that when time comes apart, the present paradigm comes apart. So there has been a lot of 'doublethink', with the aim of shoring up the present picture, and cause and effect is often ignored.

The fact is, events happen in a definite order in the universe. This is certainly true locally. The result is not an unconnected series of events, like a film that can in principle be edited into any order. There's a causal sequence, which constrains what can happen, and it needs a flow of time.

There are things in the universe that cannot possibly happen unless another thing has happened first (on Earth this is usually about filling in a form). The causal sequence we find is yet another point that suggests a flow of time, as it can't work without one.

And with cause and effect, the block time view has an unfortunate cake-and-eat-it aspect. Cause and effect exists at one level, but not at another. Many who accept the present block time view, do so without knowing it contains that, or understanding what they've accepted.

Recently a few physicists have tried to turn cause and effect around, and use the causal sequence we find to *create* a flow of time, or to weave spacetime together. Some people think spacetime arises out of the sequence. But that's simply the wrong way round - a causal sequence needs the flow of time at a deeper level, otherwise you don't get any causal sequence in the first place, with which you can get weaving.

The problem, when boiled right down, is this: current mathematical theory, although it's largely uninterpreted, is telling us to make time superficial, but the world around us is telling us to make it run deep.

Sooner or later, we're going to have to stop listening to mathematics alone. Uninterpreted mathematics can have a whole range of meanings. Sooner or later we'll have to take out the bits that are blocking the holes in the jigsaw, preventing progress. To move forward we'll need to look at the conceptual picture, and see what it's telling us.

This kind of approach always leads to new mathematics, sooner or later. But first we'll have to reverse out of a cul-de-sac, and admit we know less, before we can start to find out more.

Part 17. The best directional clue

60. A fork in the road

In the speed components picture, the time rate behaves in certain ways like a speed - one of two speeds that combine. That's a good clue, as it points in a direction. But it leads to a better one: I'm going to set out what to me is the best 'pointer clue'. But first, let's strip away some ideas that surround special relativity, which I hope I've shown there's reason to do.

When Minkowski and others first looked at the raw mathematics of special relativity, there was more than one path to choose from, and more than one version of the geometry. But in Minkowski's time there was far less reason to try unlikely looking avenues than there is now, given the problems we've had on the one he pointed us down.

To simplify things, loosely speaking, the road initially forks into two paths. On the first path you get stuck immediately. All paths that go on from there look overgrown with brambles - impassable and impossible. But the second path looks far, far better. On that one you go along for absolutely miles. Then you get stuck on that path as well!

The first path is about adding in a fourth space axis. In 1905 all dimensions were space ones, but not a few years later, when time had been thrown into the continuum. But just before reaching the fork in the road, and the choice of axes, 4D geometry would probably have meant four space axes.

This makes absolutely no sense when you experiment with it, and putting it underneath the speed components picture has an odd result: it makes it look as if all matter is constantly moving through space at the speed of light, with most of that speed happening along a hidden space axis. This unseen motion somehow corresponds to matter's motion through time.

And in fact, you might see a few things hinting at a fourth space axis. There's the right angle implied in the Pythagoras-type geometry - it's a *spatial* angle, as found between space dimensions. That right angle has always been part of what suggested time is a dimension. And yet if the fourth axis is time, then it's non-spatial. Some assume its interaction with the other dimensions could still look like a right angle in the mathematics, but the truth is, that's utterly speculative - like many areas of the standard view.

Then there's the overall speed, the speed of light. We know c is about *space*, and how fast anything can travel through it. If the parameters of space were different, c would be a different number: it's hooked into equations like the ones that give the permittivity and permeability of the vacuum. So what we know about lightspeed says that c is a tough, down-to-earth, physical thing - arising from the gritty parameters of space.

So finding c spread partly across the 'time axis' tells us something important: that it's not really a time axis at all. The part of the motion that travels on the extra axis looks gritty and physical as well, in just the same way - as if there was a space axis there, not a time axis.

So far anyway, a few things seem to click into place if we say there's a fourth space axis, with motion along it. But this path is the one that stops almost immediately. There are plenty of reasons why it looks blocked from there on. One, where is this extra space axis? Two, why is all matter moving along it, and what could that have to do with time? And above all three, *how could it possibly have anything to do with time*, as real physical motion needs time to measure it.

Time is therefore still undefined and unexplained, because you still need it as part of the description of what's going on. So this approach simply seems to push all the questions about time to a further regress, which is not helpful - quite the opposite. It adds unnecessary baggage, without solving anything to justify adding it. So because of these very discouraging early brambles, it's a path that's comparatively unexplored.

The second path looks far more promising, and in the early 20th century we set off along it. The fourth dimension isn't a space one, it's time itself. It's a different *kind* of dimension. With this there's more leeway - it's vague, and you can adjust things. Even the motion on the time axis is ambiguous: it's not real motion, like through space. So it gets left in a no man's land of existing and not existing at once (a bit reminiscent of quantum superposition). When we need this motion, it exists, and when we don't, it doesn't. This allowed us to construct a picture that seems to work, glued up with a little doublethink to hold it all together.

But a very long way further along that path, you eventually get stuck anyway. The picture becomes inconsistent - as in the problems with block time, which we only started to get to grips with in the 1960s.

And the list of assumptions is way too long. Getting time into the clothing of a dimension, with so much that's unlike the space dimensions, requires a lot of stretching. But what made us think time is a dimension in the first place, was its *connectedness* to the other dimensions. It shifts around with them. The trouble with that is, we've never understood it. And many of the things that actually resemble the space dimensions, when looked at closely, suggest a fourth space axis is somehow present.

61. What led to both paths

But here's the really important point. In either of these two versions of the geometry, what leads there is exactly the same idea. At the very heart of our reasons to take either path is just the implied overall speed of *c* through four dimensions. Along with the suggested right angle, that's what gave us *more or less everything*, including even the original idea that all of this is to do with the geometry of dimensions at all.

And yet nothing - in either version - tells us why this overall speed should be there. It just looks very significant, because in any situation, in all reference frames, try combining the speed through space with the speed through time, take them to be at right angles, you always get *c*. It's a crucial point, and we don't understand it. But it led to key areas of our view of the universe, even though the motion that led there was later removed.

Few mention it, but Brian Greene does: *"Special relativity declared a similar law for all motion: the combined speed of any object's motion through space and its motion through time is always precisely equal to the speed of light. At first, you may instinctively recoil from this statement since we are all used to the idea that nothing but light can travel at light speed. But that familiar idea refers solely to motion through space. We are now talking about something related, yet richer: an object's combined motion through space and time."*

One doesn't often see this mentioned, but it's even rarer to see someone say this is a large part of why we decided time is a dimension at all, and that it led to our whole view of time.

He then says: *"The key fact, Einstein discovered, is that these two kinds of motion are always complementary. When the parked car you were looking at speeds away, what really happens is that some of its light-speed motion is diverted from motion through time into motion through space, keeping their combined total unchanged. Such diversion unassailably means that the car's motion through time slows down."*

'Combined motion', means taking the motion to be in two directions at right angles. And 'diverted from motion through time into motion through space' sounds almost like the conservation of energy or something. He presents the complementarity of these two kinds of motion as a basic, fundamental fact, which is just how the universe is.

It's understandable that people think this way. Many don't see this as a clue about what's going on underneath it, and talk instead as if space and time are at the bottom of the well, as deep as it goes - so we just have to take this behaviour as their nature. It's true that this might need no explanation, but if we take the speed components picture to be fundamental, as Brian Greene does, and then look back at other ideas of ours, about illusions and emergent phenomena, they now look deeply unconvincing.

And the Brian Greene quote hints at the fourth space axis I've mentioned, in which all matter is travelling at c on a hidden space axis. He says the parked car has motion at c before it drives away: *some of its light-speed motion is diverted...*. Now to him this lightspeed motion is on the time axis (if he saw it as a space axis, that's a subversive idea, altering the axes, and I doubt if he's been 'radicalised', so far anyway). But my point about the gritty parameters of space in the last chapter shows one of the arguments for it.

But what he says still threatens the standard view. The speed components picture he mentions implies the timeline motion is at a deep level. But if so, Minkowski spacetime absolutely has to be wrong: we know that beyond any doubt from a rigorous proof. In the light of that, the fact that we haven't explained the motion yet, and have been resorting to illusions and superficial effects, says something is seriously wrong.

The speed components clue is what changed my view of the whole mystery. It's the only place in physics where this elusive motion along the time axis actually shows itself, and is seen to be hooked in with other things. It's like a rare snapshot of a mole that pops up out of the ground in the garden. And when you look at it, the closer you look, the more good reason is found to think you're looking at something real.

The spacetime picture teaches us never to look at the mole at all. Or rather, more accurately, to scare it so it says 'oh no!' and rushes back underground. And spacetime has a very odd effect - it tends to put a wall between us and the clues, by making the clues look unreal. It's a clever puzzle: if we take the time issues in a certain way, a number of vital clues then look like illusions. So we start to dismiss things we might need.

62. The most interesting clue of all

Returning to the speed components idea, it's a smallish chunk of jigsaw that holds together well, but it doesn't span the gap. It's also unclear how it goes in. We don't know why motion though time, combined with motion through space, always gives *c*. But the setup still looks neat, interestingly real, and suggests some kind of mechanism.

So why this overall speed of *c*? That question would be asked more often if it wasn't for one thing: *c* the number crops up all over the place in physics, so it's utterly unsurprising when you find it. But this isn't mathematics we're looking at, it's interpretations. And it's not just *c* the number here. If you can see beyond the mathematics into the conceptual side, as Brian Greene does, it's *c* the speed that's implied.

And that speed is thought to be the limit for information moving through the fabric of space, probably arising from the nature of the fabric of space. If the parameters of space were different, *c* would be different. So *c* as a speed is about motion through the fabric of space, not time. Finding it there, in that way, says real motion through space is happening.

So the speed components approach now starts to point down a path. If the path seems to make no sense, we just ignore that for now. Instead of a space axis (one of three interchangeable ones), plus a time axis at right angles to it, it looks like there's another space axis. If so, it's hidden, and somehow has a lot of hidden motion happening on it.

That's the avenue of thought that was left to one side, even though it pops up here and there, as it seemed to make no sense at all. But let's go back to it - to the short path, the one with the fourth space axis, and look at it again. And later on, by adding in some 21st century physics, we might get further along it than they did in 1907.

Suspending disbelief for now, as is needed for exploring further along a path like this one, and accepting a hopefully temporary uncertainty about what's going on, one can move forward. I first came along this way twenty five years ago. Although it makes no sense at all, one takes this motion along the time axis implied in the geometry to be genuine motion through space. A strange picture then emerges - it's like the one with the ping pong ball, but it's about an object that is constantly changing its speed.

Imagine a moth flitting around in the early evening sky. The path it follows is a complicated, looping, trajectory, and in three dimensions its speed changes constantly. But in four dimensions, we know (from special relativity) that its speed never changes at all. With every tiny speed change made travelling in three dimensions, a very accurate, rapid adjustment is made to its speed on the fourth axis. This exactly compensates for all the irregularities, and keeps the overall speed at *c*. Whatever they are, we know these 'adjustments' are rapid, partly from particle experiments.

This also happens in standard special relativity, and it happens whatever kind of axis the fourth one is. No-one has even tried to explain this, but it happens with all matter everywhere. The crazy dance of heated particles across the universe, wildly irregular in three dimensions, is smooth as glass in four. So it seems that things get evened out somehow, and that there's some kind of deep regularity underneath everything.

Now here's the clue that comes out of this. In this version of things, with an extra space axis, matter is behaving in an oddly familiar way. Although for some reason it adjusts itself constantly in how it's spread across the axes, it nonetheless always keeps on travelling at the same speed, the speed of light, and never stops.

And something else we know of does that - light. Electromagnetic radiation also keeps on travelling at that speed, and never stops. So you might say that *matter is doing in four dimensions what light does in three.*

And that's the clue. It's a similarity between light and matter. We find it if we take the speed components clue seriously, and listen to what it's saying to us about the axes, however unexpected that may be. Why does matter behave like light, but using an extra dimension? We know that waves in the same medium, whatever their size, all travel at the same speed. And light waves all travel at the same speed.

So because matter *also* seems to travel at that speed, even though it does it in four dimensions, it might be more like light than we think. This means it's possible that further along this path, in one way or another, there might be a link between light and matter. This chimes well with the chapter on confined light, which says the same thing.

And perhaps looking at '*matter does in four dimensions what light does in three*', we've got a direct clue about what the structure of space looks like. That's why I said it's the best pointer in a direction. It's a hint about the very thing we need to get to, to give us any chance of actually solving this puzzle: a picture of the structure of space.

63. What *is* the speed of light?

This moves the goalposts about time dilation. And moving the goalposts can help a lot, just as 'interactions, not measurements' did in Book I, on quantum mechanics. With these century old puzzles I've been struggling to solve, I find moving the goalposts is the best shot you have. The original goalposts have been kicked at for a long time.

An object can change its speed a lot in three dimensions, but it never does in four. So instead of asking why time is slowed onboard a moving object, we can ask a different question: why does the object always insist on keeping to a constant overall speed?

The new question is a wider one. The earlier question asks why adjustments are made, but the new question takes all this into a wider framework. It also chimes with other things in physics, and it's a far less explored avenue, so it's worth taking a look at it.

But at first, the only thing you can really grab onto is the speed of light. This makes you think, looking for clues, about what we know about c. Lightspeed is fundamental in the universe, but we don't know why in any complete way. It's the speed of several things that only ever travel as waves do: ripples in the electromagnetic field, gravitational waves. So waves seem very at home with it. On top of that, for some reason, it's also a speed *limit* for matter and information to travel through space.

One possibility is that lightspeed is the transmission speed for waves in space itself. This is widely believed, and evidence for it comes from many directions nowadays. It's a *very* common idea among physicists, but because it hasn't been pinned down, it's not well known elsewhere - or people are reluctant to mention it, in case it's subversive. Amazingly, a 2021 New Scientist magazine article - on why the speed of light exists - didn't even mention the possibility that lightspeed is the transmission speed of space for waves. This is so out of touch it's an extreme case, but it shows the point.

There's a long list of reasons to think c is the transmission speed for waves in space. Some of them are in Book I, but I'll mention some briefly here, as the point is essential to the theory I'm about to set out.

The transmission velocity of a material is set by its properties. The faster the medium springs back, and restores itself to its former state after a wave has passed, the faster waves will move through it. So that's why you get a fixed velocity. And waves of all sizes tend to travel through a particular material at the same speed, just as light waves do in space.

In the standard view of electromagnetism, light is thought to travel without anything vibrating except the electromagnetic field - whatever that really is. It's thought that there's no material-type transmitting medium. But on Earth, where we developed mathematics for waves, an elastic transmitting medium was underneath them. Light waves are similar mathematically, and a lot of physicists think space behaves in the same way. In fact, many did by 1950, as George Gamow wrote at the time. But it still hasn't been pinned down, and without an exact structure for space, we think in terms of fields, even though we don't really know what fields are.

General relativity also implies it. In GR, gravitational waves arise as 'waves of curvature', in the fabric of space. You have a curved indentation in space for a gravity field, and if it changes shape, any change would spread like a wave moving through space itself. And according to general relativity they travel at c, as if the fabric transmits waves at c.

And we know they exist. Contrary to some urban myths, gravitational waves were first detected 45 years ago - we've known that they exist beyond doubt since then. Whatever the waves actually *are*, that detection also confirmed their travelling speed. The binary pulsar was emitting gravitational waves at c, and the measured energy loss per second was as expected, and continued over time. That means the emission speed per second of energy away from the system was also as expected.

So although things are incomplete, it's clear that c is the transmission speed for waves in the fabric of space. And that expected speed was confirmed in 2017, after the LIGO detections of gravitational waves.

64. Where does this get us to?

If you now throw in the clue from the previous chapter, you reach the weird idea that light and matter both involve waves travelling *continually* through the fabric of space. So the next thing is to search for a lateral jump about the structure of space that allows this setup.

But there are still many rivers to cross. It looks like there's real motion on a fourth space axis, but it somehow has to correspond to motion through time. So how do we put that together? Well, go by the clues. They say use standing waves, and that matter does in four dimensions what light does in three. It sounds simple enough - what could possibly go wrong?

Section II. A possible solution

Part 18. Starting concepts

65. When sound came to the movies

It's easy to drive a car if you don't know how the engine works. People use computers without understanding them. We breed new plants and animals without knowing how some of their biological systems work.

Some areas of physics are the same. We've based most of our technology on quantum mechanics, which led to the standard model of particle physics, our main area of knowledge. But there's no agreement on how quantum theory works. We can predict what will happen in events to many decimal places, and have got amazingly good at using the theory. But the only consensus view is that we don't understand it.

Driving a car without knowing how the engine works is fine, unless it breaks down. Then you need some kind of roadside insurance, but that goes outside the analogy. In a way physics has started to break down - after making good progress for centuries, we've got to a point where we find our main theories don't fit together. And the thing about them is: they should.

So we need a manual. What I'm saying is that nowadays any serious attempt to unify physics into a picture could be worth looking at, as a picture may be needed. From here on, it might be impossible to move forward significantly without a deeper understanding.

The picture I'll set out now can only be reached via three lateral jumps. All three must be found, and because they're interdependent, each is harder to find on its own. That's what it was like from where I was standing anyway. I found them partly by stumbling onto them, and partly via reasoning, and the clues I've mentioned. Also really just by looking for so long in that area, that it becomes more likely that you find anything to be found there.

When I've set out the starting picture, I'll use it to explain some unexplained things in physics. They're all things that we know all about what they *do*, but not what they *are*. On the list are energy, inertia, time dilation (both kinds), the particles that suddenly appear in the vacuum, the fact that light's speed is the same however you're moving, $E = mc^2$ or the mass-energy equivalence, and others.

Talking of $E = mc^2$, contrary to one of the main 21st century myths, mass has not been explained. The Higgs field, and the Higgs boson, shed light on 2% of the mass in a normal object. That's the elementary particles it contains, but the 98% is unknown territory. So both mass and energy need to be explained more fully, along with their unexpected relationship.

A good explanation shows why what you're explaining does all the things it does. So if you want to explain energy, it should cover energy conservation, relativistic energy, the mass-energy equivalence, and so on.

The need for explanations has not been given priority in the past - some have assumed mathematics is all we need. But then, mathematics is all we've ever had. We've never had much explanation for the foundations, from Newton's laws of motion on.

What seemed the best explanation we ever had, for anything, was spacetime curvature. But this century we see spacetime differently. Many think general relativity is an approximation, which doesn't work in some situations. New, more accurate measuring instruments have been showing some cracks in the standard picture, including general relativity. New gravity theories have been getting a lot more attention recently, and spacetime curvature is questioned nowadays.

But because we've rarely had reliable explanations, not everyone sees how urgently we need them now. We know what a clear explanation is like for an individual system - you get to the mathematics, *and* an understanding, and it feels good.

I'm hoping to get some genuine explanation in, underneath the landscape in physics. The aim is to provide pictures. It should be like when sound came to the movies: they only had film, then sound arrived. With this, we only have equations, then pictures arrive. That's the aim, but it's early days.

Before drawing the picture, I'll fill in the background on a few concepts that go with it. For those who know of them already, please bear with me. There are not many, and they're all quite well known.

66. Kaluza-Klein theory

One day in 1919 Theodor Kaluza, a German mathematics teacher, was trying things out with the equations of general relativity. The theory was very new, and there was a lot of experimenting to be done. He tried casting the gravity equations in five dimensions instead of four, perhaps just to see what would happen. (In fact, something similar had been done by Gunnar Nordström a few years earlier in Finland.)

What came out looked like the mathematics for electromagnetism, together with Einstein's gravity equations. So it seemed he'd found a way to combine the two main theories of that time. But he had done it using five dimensions, and the problem was, we only seem to observe four. He decided there must be a fifth dimension, but couldn't understand why it was nowhere to be seen. He wrote to Einstein about it that year, who wasn't interested at first, but later encouraged him to publish what he had found, and even sent in the paper to the Prussian Academy in 1921.

But the mystery of why we don't see this fifth dimension remained. Then in 1926 a Swedish physicist, Oskar Klein, suggested an unexpected solution. He had been working in the same area, and he thought the dimension might be curled up to a closed circle, and far too small to see. This would mean that an extra *direction* in space curves around on itself, and anything moving in that direction would quickly return to its starting point. The dimension Klein put into his mathematical picture was Euclidian, so apart from being circular, it was just like any other dimension.

The good news was that they could calculate its size. The bad news was that it was ridiculously small, near what's now called the Planck scale (about 10^{-35} meters), which a century on is still the smallest scale we know about. At the time it seemed far too small to deal with at all. So it was overlooked for sixty years, remaining not much more than a mathematical curiosity.

And any initial spark of interest was diverted, as people were very focussed on another area. In 1926 when it was published, quantum theory was just being pinned down and put on the map, and one of the main breakthroughs of the century was reverberating around the physics world.

Then in 1938 Einstein published a paper with Bergmann, in which they tried to use Kaluza-Klein theory in an attempt at unification. Einstein studied it intensively for five years, until 1943. He couldn't find what he was searching for, and eventually concluded that it didn't exist. During the 1940s the theory was completed by others, and in the '60s it was shown that by adding more circular dimensions, KK theory could reproduce not only electromagnetism, but also the weak force, and the strong nuclear force. So it might do all the forces one day. This was interesting, but the theory had problems, and it still has, so to many it always remained a sort of 'space oddity'.

I'll return to Kaluza Klein theory in Chapter 95, which is about the future of physics. It was rediscovered in the late 20th century, and although KK theory itself was shelved, curled up dimensions, often making cylinders, started to be included in a lot of theories. There was now reason to add more than one - the number varies between theories. And since the '80s, when string theory took off, the mathematics of compactified spaces - small-scale dimensions, sometimes folded up together - has been used a lot.

The well-known analogy involves a straight line, which when you look closely turns out to be a thin cylinder, like a hosepipe, instead of a line with no width at all. This kind of dimension is too small to be reached by direct experiment, but there might be indirect effects that can be measured.

And it's thought that very soon after the universe began, a number of these extra dimensions curled up rapidly like rolls of wallpaper, each to a closed circle, during the very first instant, 10^{-43} seconds. Some physicists take these structures and shapes in space more literally than others - to some they're more like a mathematical device. But nowadays theories are suggesting that they're real, and they're widely taken that way.

67. Stepping into a new area

I've talked about the dimensions in the first book. But their nature matters in this one, and the way our idea of them has changed. Incidentally, the word 'dimension' has a separate meaning outside its literal meaning in science, in which it's used as a loose analogy for other ideas, from what might be called metaphysics. So like the word 'energy', the word 'dimension' has more than one meaning. If they're kept separate, and not confused, there shouldn't be any problem.

In physics a dimension is a possible direction in space, which helps us define that space. The number of normal, large-scale dimensions in any space is the number of co-ordinates needed to specify a point in it: it's two for a point on a map, three for a point in a volume space. Finding a building on a city map will take two numbers, but it's three if the address says a particular floor in the building.

Before the 20th century, we tended to think this was our own invention. The dimensions were a device cast by humans for convenience over a universe of featureless empty space. Some thought this space had basic properties, but few thought it had actual structure to it. But during the 20th century, space started to do strange things. It started to behave like a material - stretching and rippling, which it had never done before at all. And alongside that, the dimensions started to look increasingly real as well.

The mathematics of imaginary spaces with more than three dimensions had been explored before Einstein, but few thought they had much bearing on reality, as space wasn't seen as having any structure. Then Einstein brought in the idea that space can be curved at a large scale. To do that he made use of mathematics that existed already, about hypothetical 4D spaces. Twenty years later, Klein worked with curved space at a small scale, and sixty years on again, others started finding that a fruitful way to go. So nowadays we do think space has a physical structure, and one that relates to other structures we know about, made of matter. This has been a step into a completely new area - we've started taking the dimensions literally.

So it seems, surprisingly, that the imaginary lines we drew in space somehow actually exist. This isn't quite as odd as one might think, because the system of dimensions is utterly simple and logical. And it's not as if the structure we originally drew is exactly like the one we've been uncovering. We never drew circles, but it later turned out that there seem to be straight lines and very small circles as well.

As we find out more about the structures of space, the only thing we have to compare them with are the structures of matter. We don't know about much else anyway. And analogies are now being drawn between space and matter, as in 'curled up like rolls of wallpaper'. This imagery from interior decorating is part of a pattern of analogies between space and matter, which Einstein started a century ago. And analogies in physics are not always loose ways to illustrate things - they're essential to how we think, and theories often arise directly from them.

Ever since we started on this pattern, the principle has been holding up well, as we go further in that direction. The word analogy may be misleading. This means the structures of space *really are* like structures of matter, and really behave like them. The ideas that follow are about going even further in the same direction, towards taking the dimensions of space literally, and treating space as comparable with matter. The similarity is not a useful coincidence, it turns out to be there because matter is ultimately made of space.

68. Starting assumptions

I'll outline two starting assumptions for now. The set of assumptions needed for this theory are far fewer than it might seem, because there's a trade-off. A handful of basic ones are made at first, and later a much larger number can be dropped. So bear with me for now, and please temporarily assume one or two things. The first is one that I've already mentioned, which is that c is the transmission speed for waves of all sizes in the fabric of space.

The second assumption is the same as the first! But it's more specific: it's the idea that this goes right down to the Planck scale, and includes the fabric of the circular dimensions there. Space at that scale looks like parallel cylinders, all pointing in the same direction.

For one of these tubes to transmit waves along its length, it has to be able to ripple and shift in relation to the other dimensions, so it needs to change its shape locally. It briefly deforms in a particular place, then returns to its usual shape after the wave has passed by. A larger wave ripples through many of the cylinders.

But the cylinder is made of dimensions. So when a wave passes by, all that will happen in one place is that a *direction* in space is altered temporarily. To change its shape, the dimension will need properties similar to those needed for curling up, which it's thought that these dimensions did originally. So this assumption isn't too speculative, because we've been working with similar ideas since the '80s. And the many reasons to think space can transmit waves make the idea a very reasonable one. But right now I'm no longer trying to persuade you of these things - instead I'm just asking you to try assuming them anyway.

This book now becomes simpler than it has been anywhere else so far. I'll tell you my solution, and it's a very simple one. It involves straightforward but unexpected ideas. It explains all the key effects relating to time we observe, and the mathematics. But it also explains a lot more.

So the structure of this book is like Book I, which also covered a puzzle that has been hard to crack for a long time. We spend the first half hammering on the wall from one side, looking at the clues, and trying to use them to break it down. We don't do much damage, and we don't get arrested or anything, but the puzzle comes into focus. Then in the second half we go through the wall, look at it from the other side, and hopefully see what the clues were telling us all along - that part starts now.

Part 19. A starting point

69. The first lateral jump

Now we get back to the underlying puzzle, which is about matter's motion though time. The main single question is why a moment called 'now' appears to move along everywhere in the universe, and whether this moving present moment exists at all.

As I said, this is the very deepest crack in our current picture, which makes it the perfect starting point for trying to tackle some of the central problems in physics. Where the picture fails to hang together, there might be only a few ways out of that. This might seem bad, but it's not, it's good. It can help us to find our way, by limiting the possibilities.

So the universe seems to have a flow of time in it, or an apparent one, that can change its speed locally, affecting the rates of all processes by the same factor. Because interpretations work via analogies, it's worth thinking about anything else we know of, at all - that does something similar. It might sound like an odd question, but do we know of anything else that creates a 'speed adjustable' apparent flow of time, in which all processes can be sped up or slowed by the same factor?

In fact, there is a simple principle that does that. It's the way a film works. A film creates an apparent flow of time, and this flow arises from the frequent replacement of objects with similar objects.

In a film, objects move, and time seems to pass. If we ask the question 'what causes the flow of time in the film?', the answer would have to be that things on the screen are constantly being replaced by other similar things.

The objects on the screen look the same from one moment to the next, but they're not. A celluloid filmstrip has a series of frames, and although any two adjacent frames look similar, they're two different photos. One frame is not the same as the next, and an object on the screen at a given moment is soon replaced by a different object, part of a different photo. (They came from the same source, but that's irrelevant to the principle: animations do the same.) Their only true link is similarity. The rapid sequence of frames appearing on the screen gives the sense that time is passing.

This applies to all the different technologies that use the same principle, with a series of frames, such as digital video. But describing it with the simple film technology of the early 20th century makes it easier to get to grips with, and to use in an explanation.

In a film, the rate at which all the objects on the screen are being replaced is proportional to the time rate in the film. If you double the 'replacement rate' - double the projector speed - the time rate in the film will be doubled. Then the speed of *all processes* in the film will be doubled, even if there are many of them, all happening at different rates. The projector speed is an overall speed, so multiplying it by any factor - say 6.715 - will change the speeds of all the processes in the film by the same factor. It's clear the apparent flow of time in the film arises from this overall process, as there's absolutely no way of separating them. Whatever you do to the replacement process, the time rate in the film will follow suit.

That's all very well. But this principle was invented as a part of an artificial, man-made system. So the next question is whether anything like that could arise in nature. At first glance, it seems pretty much impossible.

70. Something strange about waves

When a water wave moves across the sea, nothing moves forward. Each area of water goes up when the wave arrives, then down again after it has passed by. Nothing actually moves forward, but something certainly seems to move forward, and it's what we call a wave.

We're used to seeing waves near the shore just before they break (because we live on the land), and by then a wave is carrying water along with it. But further out, without the slope of the shore underneath the surface, the only real movement is an up and then down movement of the water as the wave passes by.

This means that if you look at a wave, then look at it again a little later, it's made of different water. So in one sense it's a different wave. In those terms it has been *replaced* by another similar wave, and as it moves it is constantly being replaced by other waves. This replacement process happens at a speed proportional to the wave's speed across the sea. If for some reason the wave changes speed (perhaps due to smaller waves slowing it down), the rate at which it is being replaced will change proportionally.

When someone sends a wave along a rope, this is even more noticeable. You see what looks like forward motion, but the only real motion is an up and then down movement in the rope as the wave passes by. The wave is being replaced all the time, but not the rope it moves on. It's like a Mexican wave going around a football stadium - each person goes up and then down when the wave passes by, and you get what looks like sideways motion. The wave is constantly being replaced, but the crowd is always the same crowd.

71. The replacement principle

The football stadium actually looks a bit like the closed string in superstring theory, which is a circular loop. The way we've sometimes pictured a string, there are little waves going around its circumference. But a string is widely thought to be made of some kind of 'stuff', so you can only really apply the replacement principle to a closed string in a limited way. If there were waves travelling around one of these loops, the waves would be being replaced as they travel, but the object itself would always be the same object.

In the theory I'll show you now, there's a difference. It's called Planck scale time theory, PST. The fundamental object in this theory resembles the closed string from string theory, but unlike that object, it is not 'made of' anything. It consists only of a group of waves travelling around in a circle through the fabric of space itself - in the fabric of a circular dimension.

So this little object is the same size and shape as the closed string, which is about the same size and shape as a circular dimension. But the difference is, this object has no enduring existence in its own right. As its waves rotate, each wave is being replaced. But that's not all - the whole object is nothing but a disturbance in space, so it's being replaced as well.

So matter in its most basic form is a small knot of rotating waves. But what gives this object its nature as a bit of matter is rather fleeting - if you look at it a moment later, it's not the same object, it's a different one. It has been replaced by a similar object. And as in string theory, *everything* is made up of these fundamental objects: all the larger structures. So all the matter in the universe is constantly being replaced by other similar matter.

72. Use of the word 'time'

In this book I've used the word 'time' just as it's used in physics. It would be confusing to do anything else. Time is 'that which is measured by clocks' - in other words, something we know can vary its rate. In those terms, the theory that follows gives an explanation for how time works.

But it also suggests a deeper level, involving Newtonian time. That means a flow of time that's always the same everywhere, and never changes its rate.

According to this theory, what Einstein discovered is a more superficial level, at which the 'rate of change' of matter varies (as one might say if the terms were altered), rather than its time rate. But staying with the standard way of naming things works well, for one thing because there's hardly any need to refer to this underlying Newtonian level anyway. Most of what it does is just to provide a backdrop, allowing the main mechanism to work.

In the early 20th century, when it was found that the speed of *everything* can vary, physicists called these changes time rate changes. It made good sense to, as a simple example will show. On the roof of my house, time runs faster than in my living room, because it's further from the centre of the Earth's gravitational field. The processes that run faster up there include the regular pulsing of natural clocks such as atoms, biological processes in the birds and animals that sit up there sometimes: anything and everything.

The world runs faster up there by a factor of 1.0000000000000005464868 (if you've done the calculation: yes, alright, it's a bungalow - what's wrong with that? There's less stigma attached these days. It's rented from the man next door, his Mum used to live there). When these variations were discovered, if people hadn't called them time rate changes, they'd have had to describe things in an even worse way. They'd have had to say that time is undefined, and unaffected by these changes that happen to everything else. And then they'd have had to say that whatever varies up on the roof is also undefined, but something other than time.

It was bad enough having one thing that was undefined, it would have been impossible to have two. So partly for that reason, the word 'time' in physics is used for what clocks measure, even though they vary.

Nevertheless, over the hundred years since that discovery, some have been aware that our clocks might be measuring two things at once - an underlying steady flow of time, as in Newton's view, but with slight adjustments on top of it, as in Einstein's view, perhaps caused by some mechanism. Einstein's theory, after all, has put very slight corrections onto Newton's work in other places. If so, these slight alterations to the time rate would happen, for some reason, when objects are at different heights in a gravitational field, which happens a lot. They would also happen, for some reason, whenever objects move.

Part 20. A visual picture

73. Zooming in

Now let me draw the picture for you, and I hope it makes sense. Imagine the wide sweep of the universe, with clusters of galaxies threaded together like strings of pearls, or luminescent lines and filaments of egg-white. The mass curves around and rejoins itself, making an arched structure that looks like glowing coral caves, or Emmental cheese that has been pulled out very thin. The trails go curling around, here and there with visible points of light dotted along them, and huge voids in between. Then zoom slowly in.

If you can, try to aim for a bit of matter. We go down in scale gently, drifting through superclusters into clusters, into an individual galaxy, then on to stars and planets, rocks, atoms, then on down further, to smaller particles.

Getting down to the Planck scale at last, something unexpected now appears through the mist. If it were possible to see what the world looks like at that distant hidden scale, including the dimensions, as we slowly approach that level, what appears is an endless landscape of rotating cylinders. This is the 'engine room' of the universe, and the fine grain of space itself - it's a regular structure that's constantly in motion. The cylinders are all moving around incredibly fast. This is because the circular dimensions are rotating in relation to each other.

They've been spinning like this for a very long time. This rotation might have started immediately after the universe began, which is thought to be when the circular dimensions curled up and took their present shape - in the first 10^{-43} seconds. But whether or not they were left rotating after rapidly curling up, this rotation is happening now.

A dimension is simply a direction in space, even if it has curled up to a circle. So one way to picture this is to say that at every point in space, a particular *direction* is rotating. But the direction itself is already curved - it has taken the shape of a circle. So if it rotates, it rotates around itself. That makes it a different kind of rotation, and it's kept at a very small scale.

There's absolutely no reason why this shouldn't be happening at the Planck scale. It wouldn't show itself, and we'd have no way of knowing about it - not directly. But there might be indirect ways, and perhaps a range of clues that, once understood, would start to show us that it's happening.

74. Rotation theory

Planck scale time theory, PST, is less formally called rotation theory. It gives a broad interpretation for fundamental physics, which starts out from a simple background picture. That picture then helps to explain several existing areas of physics that have incomplete interpretations at present.

With a circular dimension - a familiar idea from string theory - every point in space has a tiny circle attached to it, which curves off into an extra direction, then comes round and back. The plane of this circle is at right angles to the other dimensions, although we can't really visualise that.

If a circular dimension was rotating, each point in space would be like a small rotating circle. Across the wide universe, this invisible 'background rotation' is happening, with spinning circles coming off from three-dimensional space at every point, making cylinders with the flat, straight line dimensions. I'd say the image below doesn't do it justice, the real world is far, far more beautiful than that - but it might give some idea.

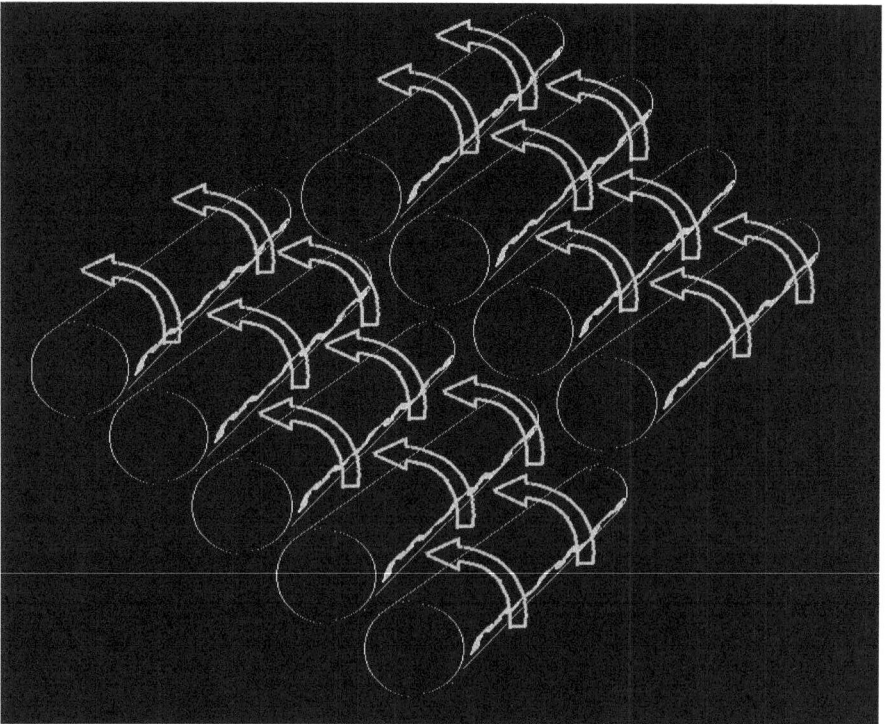

The plane of each of these circles is aligned with all the others. You could say they all face in the same direction, but it isn't any particular direction. How they're orientated depends on how the other dimensions are orientated, and that can be an arbitrary choice. So any positioning of the circles might be just as true as any other, if some basic rules are applied, because the axes of the dimensions can be positioned anywhere.

This concept shouldn't be confused with other superficially similar ideas - it's at a scale far smaller than most areas of physics. It may be hard to picture at first, but it's compatible with the way we see the dimensions. It's a sideways jump to think about a circular dimension rotating. We're used to a different kind of motion: matter moving in relation to matter, not dimensions moving in relation to dimensions. But there are hints that suggest dimensions move in relation to each other. There are also very strong clues that space behaves like matter in some ways. And one can certainly take a circular dimension to be rotating, in relation to all the other dimensions.

75. What can the matter be?

So the universe has this 'background rotation' happening across it. What we call matter is places where this rotation is slightly uneven. In certain places, the circular dimension wobbles a little as it rotates, and this creates circular patterns of waves travelling on its circumference - vibrations in the fabric of space itself. Each bit of matter looks like a small circle of waves. These waves move around the circular dimension extremely fast, as if 'running to stand still' - they're standing waves.

The dimension itself is rotating the other way at the same speed, lightspeed, which is the speed at which *anything* can push its way through the fabric of space. The transmission speed for waves in space is also the speed at which the dimensions can rotate in relation to each other, because to do that they effectively have to 'travel through' each other, which is like wave motion.

So the circular dimension travels at c in one direction, and where there are irregularities, little waves travel around its surface at c in the other direction. These disturbances arise right where dimensions join up, where one travels through another, and repeating glitches occur. So all matter might be seen as arising from flaws in a not very well-oiled system, which frankly doesn't run too well. On the other hand, the errors create something interesting. These standing patterns of waves cluster together in many kinds of arrangements, and they form into all the larger patterns and structures that make up the world.

But these structures are constantly being recreated. For a loose analogy, in rotation theory all matter is like the patterns you see in flowing water. If you look at a stream, three-dimensional shapes are visible in the flow, and each of them keeps its shape and position, although water is constantly pouring through it. Even water from a tap has shapes in its flow that are continually remade, and will keep their positions over time. These 'objects' are endlessly being replaced by other similar objects.

In this view of the universe, all matter, and the world around us, is made up of these kinds of patterns. There are no lasting things at all, just places where similar shapes are recreated again and again, making objects that appear to have duration. But that makes a world that works well, it's just not what we expect.

Returning to the main picture, we have this background rotation happening everywhere, with matter arising as places where it gets slightly irregular. It's generally thought the structure of space starts to break down at the Planck scale, though we're not sure how. It's also believed that matter originates at the same scale - this picture is a way of putting these two ideas together, and saying that matter *actually is* places where the structure of space begins to break down at the Planck scale.

It has often been assumed that matter is separate from space. Where theory has been able to go, the general picture has been that, along with radiation, the universe contains two things, matter and space.

But if matter could somehow be described as arising from vibrations in space itself, and these vibrations can hopefully stay in one place and not run away, then instead of two things there can be just one, and everything reduces to space. This allows a lot of assumptions to go - describing matter in terms of space has been a goal for some time.

Einstein wrote in 1924 that he thought matter would eventually, one day, be seen as being simply a part of space. He said: *"...According to consistent field theory, even ponderable matter, or its constituent elementary particles, are to be understood as fields of some kind, or particular 'states of space'. But it must be admitted that such a view would be premature, since, thus far, all efforts directed toward this goal have foundered. So we are effectively forced by the current state of things to distinguish between matter and aether, even though we may hope that future generations will transcend this dualistic conception..."* (By permission of Oxford University Press.)

He uses the word 'aether' here to talk about space (rather than the old idea of the aether, which his own theory had removed the need for two decades earlier). So Einstein anticipated the later general attempts to describe matter in terms of space, and he guessed that at some point in the future it might become possible.

In this picture, each bit of matter arises from irregularities in the positions of certain *directions* within the structure of space, and consists of nothing else. But the rapid rotation turns these irregularities into just the opposite of that: into something regular. It gives matter regularity, and an existence in its own right. Further up the scale matter seems to have an independent existence. And in many ways it actually has, because it has been 'spun up' into a state in which it takes on emergent characteristics, and has a nature of its own. This is why matter seems separate from space, but behaves as if it's very closely linked to it.

The rotation gives matter stability, and also causes the deep regularity in the patterns and structures it makes, such as the rich symmetries that are found in particle physics. These regularities in the many forms that matter takes are largely unexplained. They're easier to understand if they arise from a hidden regularity underneath everything, and from which all matter emerges. This regularity is the constant speed of the background rotation, and the general similarities in the way that all matter arises from it.

76. The vacuum

The outline I've set out so far can already explain one or two things that have no explanation. We know the vacuum of space isn't empty. For no apparent reason, matter will suddenly come into existence there. Fleeting particles are constantly appearing as if out of nowhere, lasting a fraction of a second, then disappearing again. The vacuum has been described as like a 'bubbling soup', with particles popping into existence all the time.

Suppose you take a sealed metal box and pump out all the air from it. Carry on emptying it, shielding it from anything and everything as best you can. It still wouldn't contain nothing at all - there'd be a vacuum, but a vacuum isn't empty. We know that short-lived particles keep on appearing there. Why this happens isn't known, but we have the mathematics that describes it. It was predicted from quantum theory, and one result of that - what's known as the Casimir effect - was later measured and confirmed.

With the Casimir effect, two metal plates, placed very near to each other, are pushed together by a force. This happens, loosely speaking, because more vacuum particles are appearing on the outside, and raining onto the outside of the plates, than in the space between them, where some wavelengths are missing, because they don't fit in there.

When that predicted force was measured, it was a real exception. No other experiment has done anything like that, and allowed us to test the space we live in. It gave us a good handle on the background behind our world, but we have no idea what's going on.

So why is the vacuum of space like a bubbling soup? If that question could be answered, it might tell us something about the nature of matter. Given that matter is actually getting started, and popping into existence everywhere in the vacuum, any explanation needs to include a way for matter to appear. So this question now edges nearer to another one: *what matter is*. That makes the vacuum an interesting area, particularly as we got to some mathematics relating to it, and it was then confirmed.

Returning to the metal box, the first point to make is that once it has been emptied, most of the things we currently know about in physics have been removed from it. And yet matter keeps on appearing in there. So this leaves not too many possible causes. All that's left in the box to cause matter, if it's caused by something we know about, are things that penetrate the shielding - some fields, some cosmic rays, neutrinos (but they're matter already) and so on.

But there's also the space in the box, and its structure, if it has one. We think it does: involving the dimensions, which we take literally nowadays. They're still in the box.

Now the dimensions are not the only things left in the box that might cause matter - some would say that certain fields are a good candidate. This is true, although we don't really know what fields are, so it might lead to a less than complete explanation. But I hoped to draw your attention to the dimensions as a possibility, being one of the few things left in the box.

Now let's think outside the box (not that one), and try to picture the circular dimensions rotating inside the vacuum box. They look like a lot of very small spinning cylinders. If you say that matter arises as brief irregularities in this motion, an explanation appears. With hidden motion going on everywhere in the box, the fleeting particles that appear as if out of nowhere can be sudden disturbances in the general rapid motion that's going on.

And as you can perhaps see, if we're looking for explanation, there aren't too many approaches that could work to explain the 'bubbling soup'. But you do have straight line dimensions, and circular ones in there. And if the circular ones are spinning, that helps get to an explanation. It makes what's in there easier to work with, and it gets a little action going in the box, to help explain how matter keeps on appearing in there.

This explanation for the creation of matter also fits the wider, more general picture. In both - whether fleeting or stable - particles are disturbances in the rotation. But it means that the 'bubbling soup', or what was measured in the Casimir effect experiment, is a bit like the spray coming off a waterfall, rather than the waterfall itself.

But we also know something else about particles generally. They have fixed lifetimes before they decay and disappear. Each type of particles has its own lifetime, and they vary - we calculate them, and some have been measured. There are long-lasting stable particles, and short-lived unstable ones, and the durations vary widely. Some lifetimes are as short as 10^{-25} seconds, while the electron lives around 10^{26} years.

Why should their durations be so predictable? If all particles are repeating disturbances in a circular flow that's happening at a regular speed, their fixed lifetimes become easier to explain. There's already regularity in the picture: a particular *kind* of disturbance in this even, regular flow will last for a specific number of revolutions. In string theory, all the different types of particles are strings, which vibrate in different ways. They're so very small that they move around a lot to create matter as we find it at the particle scale, which is much larger. In rotation theory the end result is similar, but they're different kinds of disturbances in a circular flow.

But also, *why do particles decay at all?* One very common question people ask online is why the elementary particles decay. They're not made of other components. It's easier to understand composite particles decaying: perhaps they fall apart. But with elementary ones, there has been no understanding of the fact that they decay so far.

And here's another way in which something makes better sense if one thinks of the vacuum as the background rotation. 'Quantum tunnelling' is an effect that's observed in experiments: particles sent towards a barrier occasionally pass through it. As is often the case with quantum theory, we can only work out the odds that this will happen. It's known that this can be seen in terms of a particle 'borrowing' energy from the vacuum, but it 'repays' it within a certain period of time, which depends on the size of the loan. And this kind of transaction comes up in other areas as well.

Apart from the fact that we somehow need to explain why there's energy in the vacuum to be borrowed in the first place, which can be explained by the background rotation, we need to explain these energy transactions between a particle and the vacuum that surrounds it. The close energy relationship between them suddenly makes sense if the vacuum contains a hidden flow, and the particle is a disturbance in that flow.

The odd relationship between space and matter is well-known, but surprising nevertheless. In an article on the vacuum, Roger Penrose says: *"Thus, we see that even before we need consider the mysterious effects of quantum theory, our theories of physics tell us that there is something very odd and counter-intuitive about the nature of matter. We cannot at all draw a clear dividing line between what we call 'matter' or 'substance' and what we call empty 'space' - supposedly, the voids entirely free of matter of any kind. Matter and space are not totally separate types of entity."* (By permission of Oxford University Press.)

The concepts I've described so far have been comparatively simple and few. They're rooted in ideas that are already on the map - there's no requirement for very speculative, unheard of concepts. Bertrand Russell prescribes using existing concepts in an explanation, in his version of Occam's razor. He says the more a concept is on the map already, the better for use in this way it is, because the assumptions needed will be fewer.

Although there's not much of it, the picture I've set out so far, with a few bits to be added, is enough to give new explanations for several major mysteries in physics. And it explains or sheds light on other areas, redefining concepts like energy, which have never had a proper explanation so far.

77. Four dimensions

In current physics there are thought to be four flat (straight line) dimensions: three of space, and one of time. It's also thought likely that there are some very small circular ones - different theories have different numbers of them. We only observe four dimensions - the same four that are thought to be flat. This seems reasonable, as straight line dimensions are the ones that stick out far enough to be observed.

What I'll show you now is a theory with three flat dimensions, not four, and a number of circular ones, with a backdrop of Newtonian time. All dimensions are space dimensions, and the circular ones are rotating in relation to each other. The backdrop allows them to do that. For now it doesn't matter how many circular ones there are, and we can ignore most of them - for almost all purposes you need only one of them at a time.

So what follows is a version of the theory that uses just one rotating circular dimension. Instead of the four familiar flat dimensions of general relativity, we can have three flat dimensions and one rotating circular one. The rotating one isn't the time dimension, it's a space dimension very closely associated with time. It's also, loosely speaking, a dimension that has been mistaken for the time dimension. If we say time isn't a dimension at all, the whole picture becomes simpler, and has far fewer assumptions, because all dimensions are space dimensions.

So all you need to think about is three flat dimensions, and one circular one that's spinning. All through the 20th century, people wondered what form the mysterious fourth dimension might take - it became a cultural idea as well as a physical mystery. They even wondered about it in earlier times, and frankly before relativity gave them good reason to do so.

So bear with me for now if you will, and imagine the fourth dimension as a rotating circular one. And after a while, I hope you'll hear what I eventually heard - the distant crunching sound of a lot of jigsaw pieces all clicking into position, with some of them landing in new places.

78. Matter and space

In the picture often used in string theory, a straight line, when you approach it and look closely, turns out to be a very thin cylinder. This is a surprise, but in this version of things, there's also another surprise - the cylinder turns out to be rotating.

If we take a step back, and look at the background space for rotation theory, every point in space is a rotating circle. This means that if you draw a straight line anywhere in space, a circle comes off each point on it. So the line is then a rotating cylinder. It turns out that drawing the line was like making a choice for where the dimensions are. The line you drew is now positioned along one of the flat (straight) dimensions.

Throwing away the line and keeping the cylinder, you have a cylinder made of dimensions, like the ones in some string theories. This one is rotating, and the presence of a bit of matter, if there happens to be one there, means that somewhere on the tube there's a very small circle of waves moving around its circumference.

These waves are 'running to stand still'. The circular dimension rotates in one direction, and the waves run around it in the other direction. Now being like a closed string, this object moves and vibrates in certain directions, and gives rise, perhaps, to one of the different particles we know about. But because it's not a closed string, but is instead a 'waveloop' in rotation theory, it's just a disturbance in space. However it moves in three-dimensional space, it still stays rotating on the circles that extend three-dimensional space from every point, and it's always on a cylinder like the one I've described.

That's the picture as it looks from further out, but it's easier to visualise from close up, for many situations. What you get then is even simpler: the cylinder stands still and the waves travel around it. That picture works just as well for most purposes.

How many rotations does it do each second? It's straightforward to work out that number, as long as you assume that what we call the Planck scale is at the literal Planck scale - as I said, nowadays the phrase often means a loosely defined scale. The cylinder is small, the speed is fast, but we're used to large numbers these days. Later on, when I've given more reason to see the world in this way, it might seem more amazing than it does now. Or it might seem amazing now - it certainly does to me.

The circumference of the dimension is what's called the Planck length, ℓ_P, or near it, which is 1.6162×10^{-35} meters. That's the distance the waves travel in one rotation, moving at the speed of light. So the number of times the waves go around in one second is reached by getting the number of times light can cover that distance in one second.

To get that, you need the large distance light travels in one second, divided by this small distance. That means c/ℓ_P, or the speed of light over the Planck length, which is around 1.855×10^{43} rotations per second.

That number is 1855 followed by 40 zeros. It means each second the waves travel around a circle eighteen million trillion trillion trillion times. So if this picture is right, *everything* - rocks, trees, elks, stars, wine, cheese on toast - is made of waves circling at this unimaginable rate.

79. The 'hamster and wirewheel' principle

In the standard geometry of relativity, the time axis looks like a straight line. That's what we're used to, and it behaves like a straight line, apparently with motion happening along it, at right angles to the other three dimensions. But circular motion and straight line motion can be equivalent.

Ask any hamster in a wirewheel, he knows about this principle. If you picture the wirewheel going around, you see how circular motion can 'generate' an imaginary straight line, which is a tangent to the circle.

Moving around the circle, or running to stand still as it spins, can be just like moving along this line. The imaginary straight line that is generated is similar to a real one. That's why in physics, certain things suggest the existence of a flat time axis, making a straight line, with motion along it.

This picture isn't complete until I've explained how the replacement process hooks in with it. But the wirewheel principle can already explain a few things: why time behaves like a fourth straight line dimension in some ways, but in other ways not like one at all. Although the generated straight line isn't real, the right angle in the geometry is. (The circular motion happens along a real fourth space axis, and the plane of the small circle really is at right angles to the other dimensions.) So there's a genuine right angle, which sends the line that is generated off in a fourth direction that *really is* at right angles to our three real straight lines. So what you get looks exactly like four-dimensional geometry. And in fact, it is.

And it also explains why matter's motion through time behaves like motion through space at the speed of light on a fourth axis - because matter *really is* travelling through space at the speed of light on a fourth axis, though not in a straight line. But this motion isn't a direct cause of matter's motion through time, it just has a very close relationship with the real cause, close enough to give the same mathematics.

And it's possible to show that they're proportional. The real cause of motion through time is matter's constant replacement with other matter, and here's how that process comes in.

Matter is made of waves that are rotating and being replaced. The speed of this replacement process is proportional to their rotation speed, because (as in Chapter 70), a moving ocean wave is being replaced at a rate proportional to its speed. Going on a step from there, if an object consists of a circle of moving waves in space itself, the *whole object* is also being replaced at a rate proportional to the waves' speed.

And this replacement rate for the whole object is in turn proportional to the object's time rate. That was shown with the film analogy. As in the way a film generates a flow of time, the rate at which objects are being replaced on the screen is proportional to the time rate in the film.

So that means three speeds are all proportional. They're the rotation rate of matter, the replacement rate of matter, and the time rate of matter. Each is linked to the next - the first to the second because of the wave analogy, the second to the third because of the film analogy.

But the relationship that comes out of this that really makes a difference is the link between the first and the third - *the rotation rate is proportional to the time rate*. So matter's rotation speed is closely linked to its time rate. If the circling waves in a large-scale object were to slow down for some reason, the object would be seen in slow motion.

And there's another link to be made: the rotation rate of the waves can also be related to the motion along the flat fourth axis of relativity, using another analogy - the hamster and wirewheel analogy. The circular speed is also the speed that scampers along the resulting imaginary straight line.

When you string out this chain of concepts together - all being about speeds that are proportional - from one end of it to the other you get a mechanism by which a space dimension manages to behave like a time dimension. With this mechanism in place, the fourth dimension can be a space axis, making all dimensions neatly spatial and consistent.

But because of the mechanism - which in terms of analogies is like a string of films, waves and wirewheels - it behaves like a time axis. This picture might seem complicated, but it's actually very simple. And it explains so much, that it simplifies physics in many places.

The links between these concepts also make a change to the basic geometry possible. It becomes possible to let go of the idea of time as a flat dimension, 'roll up' the fourth axis, and put a rotating circle into the geometry of space and time. This can only be done because of the backdrop of Newtonian time, which allows moving parts to be put into the geometry. But it turns out that once this change has been made, a lot more can be explained.

Part 21. Motion

80. Moving objects

So our starting point is this unfamiliar background picture, with the rotation happening everywhere. Matter arises as places where it's uneven locally, and repeating patterns of standing waves appear, and form into groups.

So if you look at a large-scale object, it will be made up of circling patterns of waves. It might be a dog in the park, waiting to run and fetch a stick, looking up to see when it will be thrown. In a few seconds the stick will fly through the air.

We know what this looks like at a large scale, perhaps not at a small one. But we know motion has something interesting about it: special relativity tells us a moving object is seen in slightly slow motion. Experiment has confirmed it, and that Einstein's simple mathematics was right. The aim now is to add in a picture underneath it, that fits the mathematics. So when the stick is thrown, what happens to these circling waves and when, about 0.3 seconds later, the dog starts to run?

The stick - like the man, the dog and the park - is also made of rapidly circling waves. The planes of the circles they make are all parallel. So the stick has a kind of grain to it (not the one in the wood). At a far smaller scale, there's an unexpected geometrical regularity underneath the world.

As soon as the stick flies, the circular paths of its waves change shape. They turn into helical paths - little 'spirals'. At a small scale the stick changes from a cloud of little circles, and moves away looking like a cloud of little springs, all travelling together in parallel.

A regular helix is a line curved as if around a cylinder, and in this case each helix actually is wrapped around a cylinder. The waves are tiny ripples in its surface. The cylinders are all parallel, and aligned with the stick's direction of motion - if this is seen from a viewpoint in which the stick is moving, and the ground is still.

As it flies through the air, the waves in the stick go 'spiralling' along, although 'spiral' is the wrong word. Their overall speed is c, as that's the transmission speed of space, including the cylinder. Before the stick moved, *all* their speed was directed around the cylinder. But as soon as the man throws it, a tiny fraction of that massive 'aroundwards' speed goes sideways, and off through the park.

Mathematically, helical motion at a constant speed gets spread between two speeds at right angles, 'alongwards' and 'aroundwards', which are combined for the overall speed. Speed components at right angles is also a way to see time dilation - that's already known. But helical motion is unexpected.

(Simple geometry explains why the two speeds in helical motion are at right angles, although one of the paths is curved. At any given moment, if you stop the film by pressing 'pause', the circular speed is moving at a tangent to the cylinder's circumference. The straight line is always at right angles to another straight line, which runs along the cylinder's length.)

And when regular helical motion keeps a constant overall speed, a particular, interesting thing happens. If more of the overall speed goes in one direction, less of it goes in the other. The faster the waves move along the cylinder, the slower they move around it. And the faster they move along it - the faster the stick flies - the more stretched out each helical path will become, and the less often it loops around the cylinder.

So this jigsaw piece begins to resemble the hole in the picture we've been looking at, because special relativity says the faster an object moves through space, the slower it moves through time. But it's more precise than that: the mathematics of special relativity, and of regular helical motion at a constant speed, are exactly the same.

81. How time slows down

Standing back, one might ask what all this has to do with time. So far in this picture, when an object moves its rotation rate slows. Hopefully that much is clear, but there's also the replacement mechanism to put in.

When the stick is thrown, you get little curling processions of waves. They're moving 'sideways' through the park, so they have further to travel for each loop they make around the cylinder.

An object's rotation rate is also the rate at which its matter is being replaced. Being standing waves, the rate of the background rotation rushing through it is what causes the replacement process. So if the rotational speed slows, the object's replacement rate slows, so its time rate slows.

Perhaps you can imagine an object made of standing waves, which suddenly moves sideways. First you have circles of waves 'running to stand still'. Then you have helixes, which go on running to stand still, but now they're shifting sideways as well, and getting replaced less often.

Returning to the scene in the park, as the stick flies, all of its waveloops are being replaced slightly less often. So the *whole object* is being replaced less often - with similar objects. The effect is like slowing down a film projector, so the frames arrive at the screen less often, replacing the images less often. The stick's time rate slows, as in special relativity. Anyone watching it fly sees it in slow motion, but they don't notice any difference. That includes the dog, who then runs, and the same thing happens again with her.

And standing further back, across the whole park there are people throwing sticks and dogs running, everywhere with moving shapes made of trillions of circles and helixes, spinning and looping around, with everything arising from slight irregularities and variations in the underlying dimensions themselves - in the crazy rippling background of freewheeling open space.

82. Any speed can be defined by an angle

There's a bit of mathematics to attach to this, which if it's not of interest can be skimmed or skipped. Special relativity and rotation theory can be brought together, so each can be seen in this situation.

Say the stick is travelling at twice walking speed, or 8 mph (3.57632 m/sec). Special relativity will give its time rate, which is hardly altered at all. As it flies through the air, although it won't be noticeable, the stick will be seen with time passing there at 0.9999999999999999288 times the rate before it was thrown. Equation 1 on page 85 will give you that number: put in $v = 3.57632$, $c = 299792458$, and calculate t, using Einstein's theory.

In rotation theory, this is also the replacement rate of the matter in the stick. The number above (with the 9s) is just below 1, because in special relativity, the time rate for a stationary object is 1. But it can also be put into units of speed like the other two speeds: multiply it by c. A significant-looking setup then appears: $v^2 + t^2 = c^2$. But it wasn't clear at all (in Chapter 44), what this was, or why t should be treated as a speed.

That number, put into units of speed and labelled 't', allows special relativity to be expressed as $v^2 + t^2 = c^2$. So we can say that because of the 'sideways' motion of the stick through the park, that speed has slowed down from c to 299792457.99999997865 m/sec, or just below c.

The problem was that although this looked significant, it wasn't clear why it should. Every physicist already knows special relativity is mathematically just like Pythagoras' triangle. But the stationary time rate is always 1, never c, so it gets a little hidden. There's a list of things that could be behind that: speed components, perspective, and other things - by finding this theory I've added helical motion to the list (under the heading of speed components).

The resulting picture not only fits the mathematical landscape, it explains it, in detail. But why is Pythagoras involved at all, the only theorem that some remember from school? That has never been explained.

Geometrically the helical path is a series of right angle triangles, one for each loop. If you could cut longways, unroll the cylinder, flatten it out, the curved line becomes a straight one. Each loop now makes the hypotenuse of a right angle triangle. A travelling object creates a row of them, like a 'sawtooth' pattern. They're identical, like little flags, but 'rolled up'. The three speeds in $v^2 + t^2 = c^2$ are proportional to the lengths of the sides of this triangle.

And some new mathematics comes out of this. Geometrically, a regular helix is based on one angle. There's a fixed angle at which the path is set, to be seen if the cylinder is unrolled. So any helical path can be defined via a single angle. And in rotation theory, that's a velocity.

So any velocity for matter, anywhere in the universe, can be defined by one angle. The angle θ is related to the velocity with:

$$\theta = \arcsin{(v/c)} .$$

(4)

An angle of zero degrees is a speed of 0 m/sec. The object has its matter on circular paths. If it starts to move, both will increase, and its units of matter all move off together, with the same angle on each helix.

Going back to the scene in the park, the picture can be improved. A moving object wouldn't look like a cloud of little springs: the angle is far too small. A stick going at 8 mph has its matter at an angle of 0.0000006835 degrees. To get a sizeable angle, say 30 degrees, it would have to fly at half the speed of light. And to get up to 45 degrees, it would have to move at 0.7071c (half the square root of 2, or $0.5^{0.5}$). We can visualise matter at that scale, but there's no way to measure the angle: probing the Planck scale directly is impossible. The theory has to be tested in other ways, and it can be.

So it's hypothetical to imagine it, but one can nevertheless think about what the world would 'look like' at that scale. And at everyday speeds, everything still looks like circles. But the waves are rotating so fast, that even travelling on a very tight helix set at a tiny angle, they'll still move slowly sideways. But rather than springs, they might look more like very thin, blurred smoke rings, moving silently along spinning tubes.

83. Light and matter

So far, bringing in the rotation theory picture only affects the interpretation. The geometry is different because it's rolled up, but the result is exactly the same. So this would be no more than an interesting interpretation, perhaps, but still indistinguishable from other ideas, and untestable.

But with this specific and economical picture in place, the interpretation can then begin to explain things, and it turns out it can explain a lot. The search can also begin for measurable differences between this picture of things and the standard one, so it can be checked - and they do exist.

Now I can describe light and matter better. ('Light' is often used to mean all electromagnetic radiation, not just visible wavelengths - it means that here.) The difference between light and matter is the direction in which they travel through the structure of space, at a scale where the dimensions make many parallel cylinders. They're different kinds of ripples in that structure.

Light travels along a straight line dimension, matter travels around a circular one. But this is more of a difference than it might seem, because the circular one is rotating very fast. So matter has to navigate the background rotation, which turns out to give it a lot of the odd qualities we know it has, but which have been hard to explain, such as inertia.

In current physics light and matter have major differences. They use different rules at times, and taking the rules for one and applying them to the other is just plain bad science. I've talked about this in the list of clues on p 91. If you look at the rules for matter light breaks (unconfined light), there's $E = mc^2$. Light has no mass, so for m, put in a zero. Now if that rule holds, E should be zero as well. But it simply isn't - even the mass-energy equivalence, however universal people would like it to be, is only universal for matter.

And light travels at a speed that matter can't reach. I've seen an article that says matter can't travel at c because its mass would be infinite, but light can, because it has no mass. So $0 \times \infty = 0$. That looks reasonable, until you think a little. Light has no mass to cause a problem by becoming infinite, that's true. But it has energy, and if it went by the rules, at c its energy would be infinite. But it's not. Light travels at c, keeping its energy small and finite.

So here's a way to show this point clearly. Two effects from special relativity are identical mathematically, and both have been confirmed by experiment - relativistic energy, and time dilation. One of them, energy, we know for sure can only be applied to matter. Light breaks the rules for it, travelling at c, but keeping a low, finite energy. So the other effect, time dilation, as it is similar to the first, *to the point of being mathematically identical*, cannot reliably be applied to light - it too can only be applied to matter.

The main thing about this issue is that answering it wrongly helps to prop up the present paradigm. So people get away with straightforward bad thinking, uncriticised by those who should guide them, and apply the rules for matter to light, in an unscientific way. But not everyone ignores the key issues, just because they don't fit our preconceptions - I've mentioned Gerard t'Hooft's paper, which looks into these questions with an open mind.

Light and matter have differences and similarities, but they're more similar in rotation theory than in standard physics. You certainly can't shoehorn them both into the rules for time dilation, but in RT both are transmitted waves in the same medium. And they travel through it at the same speed, *c*, which is the speed at which it transmits waves of all kinds and sizes. So in the picture I've been building up, light and matter are beginning to be brought together, under a single heading.

And because of that, the strongest bit of mathematical evidence for rotation theory is a near-proof that depends on the similarities. It involves applying to matter some mathematics that is normally only applied to light. And as was shown in Book II, and Appendix D here, it works surprisingly well.

84. Why does matter do in four what light does in three?

What I called 'the most interesting clue of all' needs a solution. We know an object such as a moth, which changes its speed a lot in three dimensions, still keeps a constant speed in four. Rapid adjustments somehow compensate for every change in speed, stabilising the overall speed.

It's not just the moth - all matter does this, across the universe. It happens in special relativity, and it happens whatever kind of axis you decide the fourth one is: time, space, or one disguised as the other, as in RT.

In standard special relativity the fourth axis is a time axis. You get a constant 4D speed there as well, but part of the effect is through time, so it doesn't *do* much. We've never understood it, but as I showed, it led to large areas of our present view of the universe, with Minkowski spacetime.

But take it to be a space axis, things get more interesting. This constant 4D speed, *c*, is now a real physical speed, and one that's widely seen as the basic speed for waves travelling through space. And the clue was, take it that way, and matter does in four dimensions what light does in three: it always travels at c, never stopping.

Since that chapter I've shown how light and matter travel in RT. The cylinders add an extra direction that matter uses, looping around it, but light doesn't. The extra direction makes a real difference: it allows matter to travel at c all the time. If it moves forward more, it loops around the extra axis less, and if it moves forward less, it loops around the extra axis more.

But we only see part of matter's motion. For centuries we've developed the mathematics for the visible part only. Seeing only one component still makes matter seem to travel in its own logical way - it goes by laws that have been formulated and tested, and they work. We've described matter's behaviour in three dimensions beautifully.

But it's an incomplete description: with light we see all its motion. If you add some simple extra mathematics, giving matter a constant overall speed with four axes, you'll get laws that belong in the next paradigm. I've done some of them, but RT went into a lot of different areas, and where the path split into many paths, it was hard to follow them all.

But what about the moth, what 'rapid adjustments' keep its speed constant in 4D? In standard special relativity, where time is a fourth axis, they're time rate adjustments, which happen for some unknown reason.

But using this picture with a fourth space axis instead, the rapid adjustments make sense. When the moth flies faster, the paths of the waves in its matter, like little springs, become more stretched-out. So the moth's matter is being replaced less often, and its time rate slows down. But if it flies more slowly, the 'little springs' get more compressed, so they loop around the tubes more often. So the replacement rate goes up, so the time rate goes up. Any speed change on the part of the moth is compensated for by an instant adjustment to the state of its matter. And although its visible speed will keep changing, something behind it all stays constant.

So why this constant speed, underneath all the billions of adjustments? The speed that we find everywhere in the mathematics, in all reference frames underneath everything, is really the transmission speed of space. It's also the speed of the background rotation, and the moth is a part of the background rotation. It has a constant overall speed because when you get right down to it, that's the background out of which it comes.

85. The light clock

The way a single bit of matter behaves in this picture is like something else.

It's like the old well known 'light clock', used to illustrate time dilation, which demonstrates that a moving clock ticks more slowly. The idea can be traced back to 1909, and also to Chapter 11.

A light beam is sent bouncing between two parallel mirrors. If the structure isn't moving, the light bounces vertically up and down, and the clock 'ticks' at a regular rate. Each tick is when the light hits a mirror. But if the apparatus is made to travel sideways, from the viewpoint of someone watching, the light follows a zigzag path between the mirrors. The new path is longer than the vertical one, so the clock ticks at a different regular rate, and less often when it's moving.

This reproduces the time dilation equation from special relativity, and it uses standing waves to do it. All speeds and time rates come out as they should, because of simple geometry. One 'zig' of the zigzag path is the hypotenuse of a right angle triangle, longer than the vertical path via a familiar relationship. The zigzag path creates a set of these triangles, half of them reversed. When the clock moves, the light has to cover this longer distance between one tick and the next, which slows time down as seen onboard the weird apparatus, for anyone watching it go by.

You never see them go by, because no-one has ever made one. But we know that if someone did, and if it was quite well made, it would illustrate special relativity. But that's all people really expect it to do - it's not expected to tell us anything else.

The underlying principle works because the light beam keeps the same speed from any viewpoint, or frame, which is what led Einstein to the time changes in the first place. The light clock is one of the few places where the simplicity of Pythagoras' formula is clearly visible, even to non-physicists, at the heart of special relativity. Elsewhere, this embarrassingly simple aspect of SR has been underemphasised, as we don't know what it is.

In rotation theory a single unit of matter is like a miniature light clock, and it does something similar. If stationary, it's on a circular path, but when moving it's on a helical one. For any speed, the helical path is longer than the circular one by the same factor as the one the light clock uses, with its diagonal to vertical ratio. So the two mechanisms are equivalent, and both lead to the same end result, and the same mathematics.

This means a waveloop - a basic unit of matter - is a bit like a light clock that has been 'rolled up'. In fact it's more complicated, because to move through time, it constantly replaces itself. But there's still a similarity between these two objects: the clock uses light, a waveloop uses Planck scale waves that are very much like light.

And with both mechanisms the time dilation equation falls out naturally. But there's another difference: the light clock is a hypothetical man-made object, but the waveloop is a theoretical natural object, from a comparatively small group of assumptions about the nature of space.

So in this picture, any particle, or bit of matter, is a clock that just keeps on ticking. Each tick is one rotation, and when the clock isn't being moved, the ticks come very often, once every $1 / (1.855 \times 10^{43})$ seconds. And just like the light clock, these little clocks slow down if you move them. That's because each has internal structure, and one that includes moving parts.

One interesting thing about this approach is that the time dilation equation is really describing the proportions between distances that are unimaginably small. Like the ratio, for instance, between the circumference of a circle that measures something like 10^{-33} centimeters, and one loop of a helix that's a little longer. But what makes these distances relevant to us is that the result is amplified enormously when a large-scale object moves, as the proportions between the distances stay the same.

A moving object makes trillions of loops on the helical path each second, and the distances get added together. But this still keeps the proportions exactly as they are across one loop, in an extremely accurate way.

The faster an object travels, the more extended the helical paths become. If it moves at near lightspeed, its waves would be on very stretched-out helical paths. And if an object could actually accelerate all the way up, and reach the speed of light - which we know is generally impossible, as a fundamental law - the helical paths would have to become straight lines.

86. The velocity limit

We have no explanation for the fact that the universe has a speed limit. We can pick out elements of special relativity, and point out - for instance - that it would take infinite energy to accelerate an object up to the speed of light. We know that as an object approaches that speed, the curve on the graph showing the amount of energy needed goes up towards vertical. To get the object up from $0.999c$ to $0.9991c$ takes far more energy than a similar speed increase further down the scale.

We don't have any interpretation for this, we just know that it's true. Theory predicted it, experiment confirmed it. Even seventy years ago, in the 1950s, we had large machines that could accelerate a particle up to near lightspeed, so we know this from trying. The fact is, up at speeds like that, a particle gets harder and harder to push any faster.

This is one of the places where the 'elephant in the room' in physics is even more noticeable than usual. It's the obvious fact (not taught in college), that special relativity is unexplained, and requires an explanation. The view that's often implied is 'we know what there is to know already'. But in this area we find a physical effect - we're pushing against something.

In Newton's picture of the world, you could accelerate a particle up to any speed you wanted. Increasing its speed was no harder further up, as anyone would reasonably have expected before relativity. But there's this increasing resistance as you approach lightspeed - so why this difficulty pushing matter through space? What are we pushing against?

Most would agree that it's about the properties of space. In standard physics space has properties like permittivity and permeability, which are to do with how space allows electromagnetic fields to travel through it. Mathematically, these quantities are closely related to the speed of light, but they don't tell us much about the behaviour of matter.

Light travels exactly as if c was the transmission speed of space *for waves*. So do other waves, like gravitational waves, and c looks like a wave transmission speed. That's all fine, it works well. But this straightforward behaviour has no 'sliding scale', as we find with matter. So the fact that it gets harder to push matter through space looks like something that's just to do with matter, and matter's mass and energy at high velocities.

But a velocity doesn't even exist! In standard physics, remember, a velocity is nothing but a viewpoint on how matter is moving. From another viewpoint, it'll be standing still. But a velocity is somehow real enough to make matter hard to push through space. This is why I've said the fact that it's a puzzle needing an answer should be taught to students, so they search for one. But instead it's swept under the carpet. Current physics has what boils down to one explanation for this: '*that's just the way it is*'.

Whatever this effect is, it's frame-dependent. In rotation theory frames are more real than in standard physics (as in Book I). So why does it get hard to push matter through space? It's because matter isn't loose and free to move just any old how. Instead it's 'threaded on' the dimensional cylinders, and its behaviour depends on that. And up near lightspeed, in this picture, matter starts to come up against increasing resistance.

To put it simply, this is because it's pushing against the background rotation.

The background rotation makes a sideways force, like a crosswind, at ninety degrees to the direction of motion. And the faster matter travels, the more it has to push against it.

The background rotation is part of the dimensions, and its direction depends on the frame - it's always at 90 degrees to matter's motion. That's because the rotating cylinder that each unit of matter moves along is aligned with its motion. This fits the mathematics well, and the concepts.

When matter travels at high speeds, the helical paths get very stretched out. Most of the motion is happening along the cylinder, rather than around it. To travel that way a waveloop has to 'push sideways' against the strong flow of the background rotation, and it takes more and more energy as the speed goes up, and the helical path gets more stretched out.

And if an object could reach lightspeed, the paths of its waves would stop looping around the cylinders, and would travel in straight lines along them. If travelling like that were possible, all the overall speed would be directed into the 'forwards' speed component - the 'aroundwards' component would be zero. This can't happen, because the resistance to the background rotation would have to be infinite. To push matter in a way that *completely* overrides that sideways push would take infinite energy.

It can be overridden partially, and everyone has to do that to travel at all. But the infinite push that special relativity tells us is required for travelling at c, is - in rotation theory - about overriding it completely.

I'll try to show why. The helical path angle is arcsine (v/c). In an object's rest frame, when it's not moving, the velocity is zero, so $v = 0$. A waveloop within the object is on a circular path, with a path angle of 0 degrees. If the object moves away, the path angle increases, and so does the resistance. It goes on increasing all the way up to the hypothetical far end of the scale, where $v = c$, and the path angle would have to be arcsine (c/c), or arcsine (1), 90 degrees. So you'd need a straight line at right angles to the flow.

This leads to an explanation for why the extreme of $v = c$ is not reached, and for why all the elements of that hypothetical extreme are *exactly* as they are, with some quantities going to zero, others to infinity. In rotation theory the object's resistance goes to infinity, and its replacement rate goes to zero. In relativity its energy goes to infinity, and its time rate goes to zero. A more complete version of this is set out in the chapter on relativistic mass.

The resulting setup reproduces the mathematics, and gives a detailed, cross corroborated explanation for the whole landscape. Various quantities shift in exactly the right ways, so it interprets special relativity in this area, and also explains what we find from measurements.

87. A visual analogy

I'll try to give a loose visual analogy. Imagine a river, with little waves under a bridge. They're disturbances in the current that stay where they are: ripples that run upstream to stand still.

Now try blowing them across the river, so they're moving across the flow. It's impossible, because the flow is constantly travelling through them. It's their transmitting medium, and they can't ignore that, as if they somehow existed independently of it. The equivalent of travelling at c is moving across the river at right angles to the flow, unaffected by it. It's impossible: this helps to show that matter is standing waves.

The waves on the river are a little different, but they're also running to stand still, which means they have to push upstream, at least a bit. They can travel sideways across the river, but they have to point upstream as they go. If the direction of their push swings around towards 90°, they'll be more and more vulnerable to the downstream current. So they aim a bit upstream.

But with matter, 'aiming upstream' means travelling below c. If some of the total speed is directed around the cylinder, it can't all be directed along it.

Try to push matter up to a speed of c, the energy required increases towards infinity. But there's no problem getting the waves up to c in four dimensions: three flat, one circular. In four, that's their everyday speed. But it's an overall speed, and of the two speeds it contains, only one is *visible*. What can't be done is to get that visible speed up to c.

A quick note about quantum gravity, as an aside. In the gravity book, I didn't mention this, but people are asking 'is gravity quantum?' a lot these days. In PSG it's not that simple, but loosely speaking, it seems not. Galileo found out that in projectile motion, the radial force is unconnected with any sideways motion an object has. In PSG it seems that where gravity is involved, locally the dimensions already have implied positions - in the radial direction. This also helps to explain why the near-proof from PSG is universal: it's as if even when moving sideways, an object is 'angled' towards the centre of the Earth - the small-scale behaviour of the dimensions is unintuitive but consistent.

Part 22. Energy, inertia and mass

88. Energy: questions and clues

Energy, inertia and mass are sometimes taken for granted. They're very basic phenomena, and some don't think in terms of needing explanation for them. But it's surprising how little we know about them, and whatever they really are, underneath the provisional, incomplete ideas we have about them right now, they're very important in the universe.

Before I give three simple, unexpected explanations for these things, I should set out a few gaps in our knowledge of them. I've talked to people who, due to reading from the wrong sources, genuinely think there are not many gaps in our knowledge at all. So I'll quickly set out the questions, and the answers that follow will make better sense. I'll show how incredibly unexplained they all are, but with energy I won't say much, as it's in Chapter 8.

Energy, inertia and mass are closely connected, and often proportional. We don't know what any of them are, but with mass we have some better clues now, since the discovery of the Higgs field.

The three must be taken in the right order, and energy comes first. According to this picture it's the most fundamental, and once the explanation for that is in place, all the other explanations stem from it. Energy is a universal, which applies to everything. We know of no exceptions to the rules for energy. But inertia and mass are usually about matter, and more limited - so they come second and third.

At present, there's no complete definition for energy. 'The ability to do work' is perhaps the best phrase we have, but it doesn't say enough. It says more about what energy does, than what it is. At first glance there might seem no need to say more than this.

But then you look at the tight set of mathematical rules, and the conceptual things that surely need explaining: the conservation of energy, matter's own internal energy, and the mass-energy equivalence. Einstein said that energy and mass are '*but different manifestations of the same thing*'. This applies to all matter, in any state at all.

Then there's relativistic energy: an object's energy increases towards infinity as it approaches lightspeed, even though velocities are not thought to exist, except as viewpoints on how objects are moving. Then there's that common element that links all different kinds of energy (Chapter 8), which must exist, but we have no idea what it is. And it becomes clear that there's a lot more to be known about energy than we know.

89. Inertia: questions and clues

Secondly we get to inertia. For now I'll just give a brief outline of some of the questions, which are mentioned in Book II. Inertia is the resistance an object has to changes in its state of motion. Newton thought his first law of motion was describing something internal: he called it 'the *innate* force possessed by an object, which resists changes in motion'. Nowadays Mach's principle says it's external, from the gravity fields of the distant galaxies. In rotation theory it's both internal and external, and a local effect.

All objects resist changes to their motion as if something is holding them in place. The resistance is in proportion to their mass. We don't know what it is, but when an object travels around a curve, it feels inertial forces, and it may move accordingly: it tries to keep travelling in a straight line. So how does it 'know how to move'? The idea that it somehow knows instantly has serious problems, but then so does the idea that the information takes time to reach the Earth.

And it's not even clear if inertia is a long-range effect or a local one. Mach's principle is the best explanation we have. It says inertia is a *very* long-range effect, coming from the distant galaxies. This was questioned throughout the 20th century, and since the '90s there have been attempts to replace it with a local, short-range explanation. It's that the particles that constantly appear in the vacuum slow moving objects down, particularly when they try to change direction. These ideas have only worked in a limited way, but they hint that we may end up with short-range solution.

Other things are starting to look short-range as well. We have a short-range explanation for one kind of mass: the Higgs field, which is a very local effect. But with inertia, by far the most popular explanation is still Mach's principle, so there's a new problem now, explaining the *very* close relationship of mass and inertia. The present view makes one a short-range effect, and the other an extremely long-range one. But they sometimes behave as one thing, and the mismatch may not last long: our general view has been shifting towards seeing these things as small-scale and short-range. In this general shift, mass may have temporarily got ahead of inertia.

In the explanations I'll set out now, all three phenomena - energy, inertia and mass - are short range, local effects. And the extremely close links between them are genuine similarities, rather than apparent ones. As of 2012, physics officially takes mass and inertia to be effects at enormously different scales. But the truth is, many physicists don't actually believe that, because of their ultra-close relationship.

Newton found a simple law, in which the more mass an object has, the larger the force that will be needed to create the *same* acceleration, when giving it a continuing push. Out in space, find a small rock and a large rock. Try to give both the same acceleration, from a speed of zero and increasing. Each would need a constant force applied, proportional to its mass.

This just means 'the larger rock needs a larger force'. It might seem to border on the obvious, but it's not understood. We don't really know why the large rock resists changes to its motion more than the small rock. The explanation we have has problems, and in a conference in 1993, it was found that there were many versions of Mach's principle, some contradicting others. But the idea is that the two rocks need different forces because they're affected by the gravity fields of all the far flung galaxies - which incidentally may be full of creatures also studying these puzzles.

But it's clear enough that the big rock resists a push more than the little rock because it contains more particles. Double the *volume* of some object using matter of the same composition, you'll double its particles. You'll also double its mass, energy, and inertia. So this quickly boils down to questions about the individual particles in the object.

So inertia looks local. But when you do this, and double a rock - in standard physics although you double its mass, energy and inertia, only the first two are doubled locally. Inertia is thought to be doubled across millions of light years.

There's another general point about inertia - it's to do with a question that's asked a lot less often than it might be. Inertia is about matter's resistance to changes in its state of motion, that is, changes to its motion in a straight line at a constant speed. So it's partly about matter's tendency to keep moving in a straight line at a constant speed.

The question is, why is matter so 'at home' travelling in a straight line? Why does a discus released by spinning person fly off straight, on a tangent to the circle made by her arm - and not, for instance, on a curved path? And why is that kind of motion so basic that it continues if it's already happening?

I've mentioned how physicists are always watching out for, and wary of ideas that have false intuitions and 'clinging to intuition' underneath them. They're right to be - clinging to intuition is not good science. But being over zealous about it can make one look away from a question too quickly, and miss a key point that's sitting nearby. I've shown how this has happened in a major way with motion through time. There's also the question of why matter moves, and continues moving, in straight lines.

And the thing is, asking this question at all can seem like clinging to intuition. The worry is that it might look odd merely from unfamiliarity. That's because on Earth it doesn't happen. Instead, on Earth we have friction, air resistance, gravity. Moving objects don't keep moving for long. So we get philosophers like Aristotle, who get similar issues seriously wrong in a rather embarrassing way, because of their intuition.

Naturally enough this has made people wary. But nowadays we know exactly what objects do, and it's still odd - in another way. If more people asked why matter keeps on moving in straight lines, we might start thinking about the dimensions, which, after all, include straight lines in space. And because the dimensions suddenly started to look real a few decades ago, it's reasonable enough to say that matter's motion could be connected with them.

But of course, this question has not been totally ignored. People have taken on and tackled the question of inertial motion. It's well-known that we don't have a full explanation. James Owen Weatherall has looked into it closely in a 2012 paper. He concludes that if we take the postulates of general relativity to be an interdependent set of ideas, rather that as fundamental, separate, and capable of standing up on their own, then we have an explanation for inertial motion in the geodesic principle. But it arises from, and is dependent on, other ideas in general relativity.

This hints that general relativity is wrong. If it was right, the ideas would be able to stand up on their own. Sabine Hossenfelder says the principle of least action explains why matter moves in straight lines. But that's just mentioning something we notice in the mathematics. The principle still needs explaining. We like and need economy, but - in this particular instance - we have no idea why nature does.

There's another question: matter responds *instantly* to inertial forces. When a car goes around a curve, why do we immediately move to one side? The gravity from the distant galaxies travels at lightspeed, so it would arrive late. Or do we respond to an enormously out-of-date version of the universe's gravity field? That approach also has problems.

And if Mach's principle isn't in trouble enough already, there's a large group who say it's incompatible with general relativity. Some say the same about efforts to explain inertia as a quantum vacuum effect. I've known physicists to reject *both* explanations for that kind of reason. Anyway, in short, inertia is an unanswered question.

90. Mass: new clues

Thirdly we get to mass, where new clues have come in. Mass is partly about how much matter there is in an object. All the particles add up, but how each one behaves isn't explained. What we know is incomplete, and mass still has real mysteries attached to it. The discovery of the Higgs field in 2012, from finding the Higgs boson, leaves most of them unanswered.

Different kinds of particles have different fixed masses. Knowing a particle's mass, one can work out what it will do: there's a simple set of rules. A large-scale object's mass affects how it will move, and how it will respond to forces and events around it - which usually means to other matter.

It seems likely that there are major clues about the nature of the universe in the unanswered questions about mass. A large part of the mystery is that we don't understand the close relationships that mass has with energy, inertia and gravity. These things span a wide area. None of them is well understood - the Higgs field produces some clues, but it doesn't shed direct light on the relationship between mass and any of them.

But the Higgs field, whatever it is, has been found to exist. It's thought to fill all of space, and is sometimes pictured as like treacle, in the sense that some particles move through it more easily than others. The analogy is loose and inaccurate, but the elementary particles respond to the field in proportion to their masses, and the different responses are thought to be what gives them their masses in the first place.

Like other fields, the Higgs field should have a unique particle that goes with it. The way to tell if the field exists is to find its signature particle, and in 2012 the Higgs boson was found with the LHC (Large Hadron Collider) at CERN. It looks so like a 'standard Higgs', that some were wishing it would be different enough to yield a bit of new physics.

But this relates to only to one kind of mass: the elementary particles. They're the basic building blocks, but they make a small fraction of the total mass in an ordinary object, about 2%. The Higgs field doesn't 'explain mass' because 98% of an object's mass lives without it. Particles bind together, forming into composite particles, like the nucleus of an atom - there are quarks within the protons and neutrons. They're all held together by energy bonds, which also have mass - a lot of it. But it's not understood.

So the Higgs field can't be a full explanation at the deepest level, because it doesn't incorporate that kind of mass, and it doesn't really explain the mass of neutrinos either. But *all* mass behaves in a uniform way in other areas. Because mass is proportional to both energy and inertia, we can even leave mass out, and say the energy in an object is proportional to its inertia. We simply don't know why this is so, but the total is thought to include *a small contribution* from the Higgs field.

There's also a problem with the mass of the Higgs boson, as we have no way to reconcile it with the far larger mass scale predicted via gravity. Efforts to explain this 'hierarchy problem' have been intense ever since the 1970s, but there's still no consensus at all about it. Recent experiments were expected to show a mechanism to explain it, but they haven't.

Gian Giudice, a physicist at CERN, described the unlikely balance, which has been compared to finding a pencil standing on its point: *"The issue is that in quantum mechanics, everything influences everything else. The super-heavy gravitational states should mingle quantum mechanically with the Higgs boson, contributing huge factors to the value of its mass. Yet somehow, the Higgs boson ends up lightweight. It's as if all the gargantuan factors affecting its mass - some positive, others negative, but all dozens of digits long - have magically cancelled out, leaving an extraordinarily tiny value behind. The improbably fine-tuned cancellation of these factors seems suspicious. You think, well, there must be something else behind it."*

There's clearly more going on than we understand. Science reporting tends to emphasise the areas of certainty and progress, particularly when it's about a breakthrough. A lot of non-physicists think the Higgs boson has explained mass, and has been neatly pinned down. But in terms of our understanding, we're left with little more than we had, and ongoing problems. It seems the Higgs field may be just one aspect of a deeper explanation.

It's always possible to have a mathematical description of something without knowing what it is. According to rotation theory, the Higgs field is related to something with similar properties. The 'treacle' in the analogy, which some particles move through more easily than others, and fills all of space, is really the background rotation.

The background rotation is what gives the elementary particles their masses, because each is a different kind of disturbance in it. So each moves through it in its own individual way, and has its own interaction with the background. When mass has been defined, this explains the differences in how particles move through it, and the result explains a lot more.

So the Higgs field and the background rotation might be two aspects of the same thing. The flow might be what's going on underneath our present view. They're similar, but somewhere along the line, the two theories will either diverge, or one will turn out to emerge from the other.

What we call the Higgs field certainly exists somehow, just as what we call the electromagnetic field exists. But in both cases a far wider explanation is needed.

It's easy to show there are different layers of explanation in the world, some deeper than others, with an example. Try to explain butter: at a first level of explanation it comes from milk. But at a deeper level, it comes from a cow. Because the cow provides a deeper explanation, it can explain a wider range of things than milk can.

The masses of the elementary particles are like the butter in the analogy. The Higgs field is like the milk - an explanation, but not at the deepest level. The background rotation is like the cow - a deeper layer underneath these things. It's worth noting that in farming, the milk and the cow are different, and the only way to link them, rather than showing them to be the same thing, is to show that one emerges from the other.

Both these explanations for butter are true, and in some areas, either would do just as well. But the reality is, in trying to understand mass in the way we hope to in the current situation - we need to get to the cow.

91. An explanation for energy

That's the questions about energy, mass and inertia, here are some answers. They're all linked together, just as the questions are.

In special relativity, energy is not a side issue. On the contrary, it's one of the 'key player' concepts, along with matter, space and time. They're tied in with each other, in relationships for which we know the mathematics, but not the conceptual links. But energy is different from the others. With matter, space and time, we have some idea of what they are.

Matter's internal energy is much larger than its mass: a small mass makes a lot of energy. The number that translates between them, c^2, is around ninety million billion. Inertia doesn't come in any units, but it's at the same scale as mass. Newton's $F = ma$ law shows mass behaving as inertia.

So an object's energy is enormous, while the equivalent mass and inertia are small. This means they may be a tiny side effect of something. Whatever it is, there's something going on with a small element, and a huge one. The main effect is ninety million billion times larger than the side effect, or 9×10^{16}.

Energy has a universality that's hard to explain. It suggests that something fundamental is going on. I quoted Richard Feynman about it in Chapter 8: he says of the conservation of energy: '*it is not a description of a mechanism, or anything concrete...*'. Well, below is an alternative view, in which it *is* seen as a mechanism, and something concrete.

There has to be something that quantifies energy. The different kinds - there are about fifteen - vary widely: electrical energy, chemical, elastic, matter's internal energy, and so on. The mystery is that some thread pulls them all together, like the analogy of whisky, gin, vodka, beer and wine all containing alcohol. Whatever this common thread is, it's something precise enough to quantify them.

The reasoning behind that is in Chapter 8. It seemed to me that if there's a unifying explanation underneath all the different forms of energy - all these wide-ranging, different phenomena - it must be a *process* of some kind. This looked like another clue (along with the time rate variations) suggesting that some kind of general process is going on behind everything.

And to do anything with this clue, you really need a real flow of time, as any process does. So it was another indirect reason to take the flow of time to be real. To me it showed, with hindsight, how removing a flow of time has been a major handicap, which has blocked physicists off from getting to some of the vital ways forward.

In the picture I've been describing, a particle is a disturbance that appears in a very small-scale, very rapid, circular flow. If it's long-lived, this disturbance is stable, and just sits there (though constantly recreating itself and replacing itself) for billions of years. Other particles are unstable, and disappear.

But all particles have a *resistance* to the flow from which they come, and it's this resistance that gives matter its energy. Or rather, it's this resistance that quantifies matter's energy. Energy is usually activity of one kind or another, but what links up all the different kinds of activity, and measures their effect, is the resistance the activity creates.

Activity is a positive quantity, it's about something physical happening. So it's added to the world, if you like. But energy is quantified by what it in a sense subtracts, from something larger. So in the flow, energy gets measured off as a quantity that pushes the other way, by how much it resists.

A. Matter's internal energy

Any particle is a repeating disturbance in the flow, and its activity is what will resist it. This resistance is like other things at larger scales. Different particles have different energies, because different kinds of disturbances in the flow will resist it by different amounts.

So the activity can simply be the presence of a particle, which is itself a little bit of resistance to the flow, due to a small disturbance in it. That explains why a particle has its own internal energy. We know matter's internal energy is different from the kinds of energy we understand better, because it seems that matter somehow *is* this basic energy.

Matter is at root level a repeating disturbance in a circular flow, with waves that are 'running to stand still'. A normal object has trillions of points within it where this is happening.

This means what we call an 'object' is really a disturbance in the flow: a lamp, a squirrel, a piece of toast. We know that an object's energy, if its matter is evenly spread, is proportional to the volume of space it occupies. That means more matter = more energy, and seems obvious. But in rotation theory it's about resistance. A large rock resists the flow more than a smaller one (with the same composition), so it has more internal energy.

B. Other forms of energy

But where there's activity, this also creates extra energy, other than matter's internal kind. Any form of activity does, including a lot of different things. For instance, the motion of particles in an object causes what we call heat. This general activity increases the object's resistance to the background rotation, so it adds energy - heat energy - on top of the basic energy contained in the individual particles.

C. The conservation of energy

This picture, although just an outline, can then begin to explain other things, such as the conservation of energy. Energy can morph into different forms, keeping the total amount the same - we know that much. In rotation theory, what stays the same is the total resistance to the flow.

It stays the same in the spectacular way it does, because once the resistance is there, it becomes a self-perpetuating thing. The flow carries on rolling, but with this local pocket of resistance in it. It's like a bubble in a rolling ocean wave, which is surrounded by far stronger *containing* forces than the small ones inside it. The bubble can change its shape, but some numbers will stay the same throughout the changes.

In the same way, the resistance to the flow this pocket of resistance has will stay the same. Its energy is conserved, because the surrounding background is far stronger than the details of the processes and changes it goes through. So what's called a closed system arises comparatively easily - easily enough for us to have noticed the conservation of energy.

D. Defining energy

Energy in rotation theory is 'resistance to the background rotation, caused by any kind of activity'. This might be extended to 'resistance to the background rotation, caused by any kind of activity or force'.

But what we call 'force' may reduce to more activity at a small scale. In the standard model, three of the four forces are described via carrier particles, which convey these forces. And in each case, the predicted carrier particle has been detected - so that kind of description works extremely well. It tends to reduce force to activity, across all the areas these forces cover. So perhaps all force reduces to activity.

E. Potential energy

Thinking in that sort of way, perhaps the forces at work when a bow is pulled back before releasing an arrow - the tension along the string - just involves extra activity of particles at a small scale, zooming around within the string and the bow. Potential energy is stored within the configuration of a system. But given that energy can change easily from one form to another, keeping the amount the same, and we know it's about activity, it seems likely that at least some potential energy, perhaps all, is really just energy stored in one form of activity, ready to be converted into another.

Gravitational energy should not necessarily be taken in the standard way, if one is trying to understand the RT interpretation for energy. As is in Book II, the view of gravity from rotation theory is different.

F. How forces are directed

A lot of these larger-scale forces and forms of activity are directed in three-dimensional space, which means along the length of the cylinders, resisting the flow that travels around their circumference. Activity is directed at right angles to the flow. But resistance is *always* directed against the flow: on the same fourth axis, but in the opposite direction to the rotation.

This point includes the energy bonds that hold elementary particles together in a composite particle. They add a lot of mass to the total, which the Higgs field can't explain. In rotation theory, these bonds are forces directed along the cylinders, which resist the flow around them.

G. Light

Looking at the energy of a photon, which travels at right angles to the flow, it resists it slightly as it goes. So it makes sense that shorter wavelengths have more energy. They involve more activity, because there are more waves, so more movement. There's more going on, so more resistance.

A photon's energy is inversely proportional to its wavelength, and goes up if the wavelength goes down. So energy multiplied by wavelength is constant: $E\lambda = hc$. If E or λ goes up, the other goes down, and their product remains the same. This is known, but not explained - but rotation theory says more about what's going on, and about why Planck's constant h, a known fixed number, exists as a proportionality constant.

H. The principle of least action

There's an unexplained principle that underpins all basic physics, and can be used to reach the mathematics instead, without bringing in forces, or laws of motion. Richard Feynman said it must run deeper than all our basic laws, but no-one knows what's underneath it. It's about how matter always uses the minimum energy, whenever it does anything. It's one of the most important explanations in the book, and is in Appendix C.

92. An explanation for inertia

In standard physics, resistance already comes into an area that's not far from this. Inertia, as we see it, already contains the idea of resistance.

But it's a very different kind. Inertia is the basic resistance that an object has to changes in its state of motion, or lack of motion. It's the resistance objects have to forces, and on Earth there are a lot of them at work: people picking up objects and throwing them, animals tugging at them or jumping up and grabbing them, and other objects bumping them around. So inertia is about how objects resist what the world does to them, and it's a much smaller kind of resistance. This is the small side effect I mentioned.

The large number is energy, an object's resistance to the flow itself, which is very powerful, and arrives from a fourth direction. But in three-dimensional space, there's a gentle resistance, because objects are supported by the flow that surrounds them. It's like objects floating around in a liquid. If you push them they start moving, but they gently resist the push. Then, once moving, they keep moving.

Now let's return to the two space rocks from a few chapters ago, just before anyone starts pushing them. In rotation theory all of space contains a hidden flow, and chunks of matter are local disturbances in it. The larger rock, with the higher energy, is busy resisting the background rotation a lot. As a result, it is somewhat *held in place* by the flow. It will be harder to move it than the smaller rock, which has a lower energy, so less resistance to the background rotation, so it is less held in place by it. As a result, it shifts itself more easily if you give it a push.

Both rocks have what is like a waterfall flowing through them, and a flow of water can hold an object in place. This is particularly true if it's flowing right through it. This analogy may give an intuitive sense of what causes inertia in rotation theory. The more an object is resisting the flow, the more tightly the flow 'gets to grips' with it, and so the more it fixes it in place.

This leads to another way of putting the link between energy and inertia. An object's ability to resist the flow (its energy), is proportional to its ability to resist a sideways push in three-dimensional space, at right angles to the flow (its inertia). So these two different kinds of resistance are proportional. This makes sense, because how you're held in place from one direction affects how you're held in place - ie. what a push will do - in another direction.

But the word 'ability' might give the wrong picture. It's not so much matter's ability to resist the background flow that gives it its inertia, it's more the fact that it's actually busy resisting it.

So this is a definition for inertia, which starts from the definition for energy I've set out. The two elements of it - resistance to the flow, and the fixing of objects in three dimensions by the same flow - are proportional. So inertia increases with resistance to the background (the big rock is harder to push), so inertia increases with an object's internal energy. This conceptual picture can explain their relationship, and why they're proportional.

As you can see, we're heading towards an explanation for $E = mc^2$, and have it in view now.

93. An explanation for mass

In the relationship between energy and inertia I've set out, inertia is a small side effect. But it's a very important one, because it's about what happens in a set of directions that mean a lot to us. Objects move quite easily in three dimensions, floating and bumping around, with a slight, gentle resistance to changes in their motion. But in the fourth direction, by contrast, enormously powerful forces are at work, involving resistance trillions of times stronger. And although it's a different kind of resistance, dividing by c^2 translates from resistance in one direction to resistance in another.

If this doesn't make much sense, why, purchase an air hockey table, it's quite fun anyway. Or find images of it online: in air hockey, the puck floats around on hidden jets of air that come up through holes in the playing surface. The jets come from an extra direction, at right angles to the two directions that count - in air hockey. In this case it's a third direction, while in the universe the extra direction is a fourth one. But the point of the analogy is: to describe or explain the behaviour of the puck in a complete way, you'd need to know about the action in this extra direction.

This general setup, in which classical mechanics unexpectedly turns out to be a bit like air hockey, leads to mass and inertia. There's a need to distinguish between them. Both are a small side effect of matter's energy, and the side effect is relevant to many different interactions that matter goes through in three dimensions. Objects go floating and bumping around anywhere in our three-dimensional world, and interact in different ways - with other matter, or with light.

What we call 'inertia' is a very noticeable effect in this group of them, related to matter's motion, and resistance to changes. Inertia is the direct version of the effect, but the wider term in all these interactions is mass. Mass covers far more aspects of matter's behaviour, and the mass term enters into many equations that describe matter's behaviour, but always in three dimensions. Mass is the *measurable* aspect of inertia, just as time is a measurable aspect of change. But the basic form of mass - the one that shows the link to energy (via resistance), and so shows where mass comes from - is inertia.

It then becomes possible to explain something that we very much need to explain: why the energy bonds that hold a composite particle together have mass. This is the 98% of the mass in an ordinary object, which the Higgs field can't explain. One reason there's an urgent need to explain it (apart from the fact that the public have been told mass has already been explained, so we should try to catch up, and make it true), is to pull the different parts of our overall picture together.

A composite particle, like the nucleus of an atom, is a few indivisible particles stuck together, and there can be a lot of them. They each have a mass, but the total mass is not arrived at by simply adding them up. You also have the forces that hold them together - the bonds have energy, and the energy has mass. That creates a lot of extra mass. It's most of the mass in an ordinary object, and it comes into calculations we do about particles - you can't leave it out, if you want to get the right answer.

It's one of many examples of the mass-energy equivalence being completely uninterpreted. So what does rotation theory say about this? The bonds that hold particles together do so *across three-dimensional space*. That can seem unremarkable, but in RT the result is that any pull between bits of matter is directed along the length of the cylinders. The force probably takes the form of activity, and we think these forces are conveyed by 'carrier' particles.

In rotation theory there are strong hints that these forces involve standing waves in one way or another, because the general rule in RT is: if it has mass, it has standing waves.

But whatever form the force takes, what makes it count as energy is the fact that being directed along the length of the cylinders, it resists the flow that travels around them. And this resistance to the flow also translates to inertia, and adds to the total mass of any object.

It has been hard to find a viable explanation for this kind of mass, particularly because it needs to dovetail in neatly with four or five other explanations, for other phenomena. That's a bit of a challenge, but it works in that way here. As in the explanation for quantum mechanics in DQM, for certain elements (like the way light or matter can be in a superposition), there aren't too many other 'thinkable of' ways to explain it.

94. Mass and energy, like water and steam: $E = mc^2$

The mass-energy equivalence isn't only unexplained, it's totally uncharted in the way quantum theory is, without any notion of what might be going on at all. In fact, quantum theory has a little more: a few incomplete explanations, mostly far-fetched ones, and no consensus. $E = mc^2$ has not even that, but it must reveal something utterly fundamental about our world, involving both mass and energy.

We've known about this link for a century, entire books have been written about it, and they sometimes tell you that an explanation for $E = mc^2$ comes out of standard physics. But as all honest physicists are well aware, the true answer to what's behind that equation is 'we don't know'.

Einstein showed in 1905 that energy has mass, in a separate paper from his special relativity paper that year. The paper was called *Does the inertia of a body depend upon its energy content?*, and included the sentence 'If a body gives off energy L in the form of radiation, its mass diminishes by L/c^2 '. That means $m = L/c^2$, which in the more recent convention is $E = mc^2$.

Two years later he pointed out that it applies the other way round - mass has energy, with c^2 as the permanent factor difference between them. There has never been a widely accepted explanation for this connection, nor has there really been a not widely accepted one. But in 2005 it was tested to an even higher accuracy than before.

Because matter is made of atoms, and each atom is made of smaller particles that stick to each other or fly around each other, it's not so surprising that all matter has energy. What *is* surprising is that an object's total energy divided by c^2 always gives its mass, even if there are several different kinds of energy in the object. This tells us that there's something very fundamental going on, and whatever it is, we simply don't know about it.

The link has to be very close. Einstein, who I'll quote on this again (having got permission to from a film company that interviewed him in 1948) said energy and mass are *but different manifestations of the same thing*. My partner Jill said 'that means like water and steam'.

In rotation theory, they're just that: two different versions of the same thing. They're slightly further apart than water and steam, otherwise the link would have been found. Mass and energy are measures of matter's resistance to forces, in two different directions in the structure of space. Water and steam are the same substance at two temperature ranges.

Having set out separate explanations for mass and energy, perhaps the link between them becomes clear. I've hopefully shown why they're proportional - Chapter 92 shows how mass, as well as inertia, links to energy. So whether or not one agrees with this explanation, at least it is one, and in the past we haven't had any of those to choose from.

And here's an 'out of the box' train of thought that I didn't find - I don't know if anyone else saw it. I got there via a slow roundabout route, and then saw it with hindsight. But the train of thought, which is a very reasonable one, goes like this: energy and mass must somehow be *very* close. What do we know about their deep nature? We only know one thing about one of them. Mass is to do with resistance to forces, as it's very close to inertia. So could energy possibly, somehow, be related to resistance to forces?

That's a good question, but it's too far out of the box. Perhaps some thought about it, but couldn't go anywhere with it. Looking back with hindsight later, it seemed amazing that I had missed it. Even when getting near it, in the four years or so between the idea of the background rotation, and the idea that energy is resistance to it, I *still* didn't find it - my ideas about energy were still immersed in a very different way of seeing things.

Finding this picture felt like getting into a parallel version of the world, but as you can see, I wasn't quick to react when I got there. Two decades of looking around slowly was needed to find my way, step by step, past *'the deposit of prejudices laid down before the age of eighteen'*, which is what Einstein said, or is said to have said, is what we call common sense.

95. Guessing at the future

Hopefully you now have an understanding of how this explanation works, so before we get to relativistic mass and energy (which are the only things left), I can take a break from laying out this set of explanations, and stand back to have a look at where it might go.

If this approach is right, it implies that the final set of laws for the behaviour of matter - or the next set of laws - will be about the Planck scale. The ideas in the theory push the laws to a further regress, where the large-scale rules we have arise from another set of rules. The two are often identical, for most practical purposes. But there are still differences, and I've searched for them over many years, finding one every now and then.

So thinking about future work on the theory, the questions include how the surface of the dimension behaves at that scale, shifting around enough for standing waves in its fabric to appear and keep going. The need is to get to a better description of the background rotation.

The particles in the standard model have a long list of parameters that are at present unexplained, and have to be taken as they are, without questioning them. They're adjustable in the standard model, and so far we put them in by measuring them. An enormously important goal in current physics is to explain just where those numbers came from, and we've been having a lot of trouble with that. That would be where physics really begins to connect with reality. So far, *not one single particle* nowadays thought to be fundamental, has ever had its mass or coupling constant derived from any fundamental theory. But people keep trying.

I think those numbers will be explained by using the best handle we have on the world at that scale - by taking the elementary particles to be the *stable* disturbances that can exist within the flow. These stable particles get into a repeating pattern, and just carry on for millions of years.

Maybe we'll get to a good description of the background rotation by working backwards from those numbers, trying for a version of the setup from which the same numbers come back out again, representing the stable patterns. I once told this to Neil Turok, who didn't seem too sure it would happen, but he expressed approval of the idea - of the mathematical approach.

This is not all that dissimilar to the energy levels in the atom, found a century ago. The stable electron orbits each had a whole number of wavelengths and (although in that instance it wasn't the true picture), it helped people get to the mathematics at the time.

But if this picture is right, what about unstable disturbances? We know there are also particles that - in rotation theory - fail to get into a repeating pattern for long, like the fleeting vacuum particles. And muons: stable or unstable? It depends what point of view you take. They last for 2.2 millionths of a second, which to us is nothing. But in the background rotation's terms, they rotate a lot before they disappear, 4×10^{37} times, or a 4 with 37 zeros. So in another way of seeing it, they do settle into a bit of a pattern.

Although these two worlds are very far apart, the picture I've given is quite easy to take onboard. It seems that what happens at the Planck scale is oddly like what happens in our large-scale world. That might be surprising, because the small-scale world I've described is strange - it's made only of space, and irregularities in space. But the similarities exist because the large-scale world is made of exactly the same thing, and nothing else.

Still guessing at the future, and it's a thankless task, in Book I there's a list of different kinds of lateral jumps. One of them was taking an idea that seems not to work, and altering it until it does. In Book II, refractive medium gravity was shown to work extremely well, but only when the background concepts are altered in a particular way, to fix some problems. And here in Book III, I'll tell you that Kaluza-Klein theory, although shelved, when reupholstered in the same sort of way, may be the future of physics.

KK theory has come a long way since 1926, but one of its problems was that the radius of the extra circular dimension is unstable. If perturbed it can start expanding: we don't observe that. In RT the picture is similar, but it's also spinning at c, which can resolve the problem. Other problems might also go, because they arose from assumptions made in the 20th century, coming out of standard physics such as general relativity, which people were absolutely certain were right, but which were quite simply wrong.

So KK theory becomes interesting, particularly in the light of rotation theory. It turns out it can reproduce three of the four forces, by adding more circular dimensions: one to cause electromagnetism, two more for the weak force, another four for the strong force. A large-scale dimension was thought to be needed for time, but RT creates time and gravity (the last of the four forces), without adding any more dimensions.

This is a major difference, and a game changer: it changes the landscape. So although (for reasons in Chapter 122), this version of RT works very well just as it is, perhaps that points out a way forward. The path it indicates may be a path well worth exploring. Either way, on that path, and on any of the many paths leading off from here, or for that matter, leading off from anywhere at all - good luck on the road.

96. Frames - and why relativistic mass can't be ignored

There are different kinds of mass. The most basic is inertial mass, sometimes called rest mass, invariant mass, or just 'mass'. Each type of particle, and any chunk of matter, has its own fixed mass value for when it's not moving. Even at rest, with no forces at work on it, matter stores what its behaviour would be if there were forces at work.

Whether or not an object moves, it keeps this rest mass value. Basic special relativity tells us that if it moves, that number gets combined with another. A moving object's mass is two numbers multiplied together: a fixed one, rest mass, and a varying one, relativistic mass. If it starts to move, the rest mass is multiplied by a number calculated from the speed.

Energy does the same. An object's energy also has an absolute component and a relative one. Total energy keeps in step with total mass, and as always, everywhere, it's c^2 times larger.

A reference frame is about how we take an object to be moving. It seems no more than a viewpoint - we choose it. Because mass depends partly on the frame, an object's mass depends on how you look at it. This clue is so weird that a group of physicists tried to remove it from the picture. They didn't try to change relativistic energy though.

But if you tidy up the picture you might throw out the best clues. The picture isn't meant to be complete - loose ends lying around can help. And alongside the healthy drive to work on an unfinished picture, and leave room for ideas, there's always the unhealthy drive to make the picture seem more complete than it is, and sweep the problems under the carpet. Relativistic mass is so counterintuitive that a discussion broke out about whether it should be kept at all. It was defended very strongly, and it's widely used now as it has been throughout.

One excellent point that has been made in defence of relativistic mass is that to let go of the concept, we'd have to let go of the much needed idea that an object's mass is proportional to its weight. When an object is heated, all the particles in it move around more rapidly. This causes a slight relativistic mass increase, so the object's weight increases. You can calculate it - one day we'll measure it. The change is small, but when it happens, we urgently need to be able to say the object's weight and mass have both increased. Otherwise the picture comes apart in another place instead, and in a worse way. This shows relativistic mass is real, and not to be removed.

And there's another reason why throwing it out is a bad idea: it might be the clue of clues. There are very few areas where attempts to interpret two vital theories, special relativity and quantum mechanics, overlap at all. But there's one: relational quantum mechanics, RQM. And this clue - matter's properties that vary with the viewpoint - is right at the centre of the overlap. So it may point straight at the underlying picture.

And this takes us to a key point about reference frames. Rotation theory and standard physics disagree about frames, and in a major way. Frames seem to be viewpoints on matter's motion, with no real existence at all. The standard view of special relativity, which often fails to admit that there's more to be understood, implies frames have nothing weird or unexplained about them. A velocity disappears if you change frames, and it simply doesn't exist after the change. So far so good.

Most of us agree on relativistic energy - it's not disputed like relativistic mass is. So we *know* matter's properties can vary with the viewpoint. But if you accelerate an object up to near lightspeed (Chapter 86), relativistic energy is so amazingly real, we find we're pushing against something very strong and physical. In another frame, this strong force just disappears. So there's more to frames than we realise. This is an enormous clue, but it's so weird, some will ignore it, or try to throw it out.

In RT this frame-dependence of one part of an object's mass says something. It says mass is something to do with the background rotation. The direction of the background rotation is also frame-dependent, because it's part of the dimensions. So deciding about how an object is moving affects the implied positions of the axes. So it affects the direction of the hidden flow, which also depends on their positions. And choosing how an object is moving also specifies the amount the object is pushing against the background rotation, in order to move in that way. This explains that powerful force we find we're pushing against in certain frames.

It might seem strange that the positions of the dimensions can jump around, depending on how we see things. But that's exactly it, it's about how we see things. When we pick a frame, it seems we're choosing just, say, A and B. But we might be choosing A, B, C and D, for all we know.

And frames have been shown by experiment to make an actual difference to the world - with relativistic energy, but also with time. Two frames can leave two clocks out of sync when brought back together. So something *real* that we don't understand has to be happening with frames. This means the idea that choosing a frame involves choosing a few other things as well, becomes what might be called an utterly reasonable idea.

97. An explanation for relativistic mass and energy

The explanation for relativistic mass and energy will be understood already to some. It's implied in some things we've already looked at, like the velocity limit, so this chapter is short.

According to special relativity, as an object approaches c, its mass and energy both head towards infinity. In rotation theory this is because to move at that speed, it's having to push increasingly against the sideways 'crosswind' of the background flow. That increases its energy, because its energy is resistance to the flow. Resisting the flow also increases the object's mass, by the same factor. The two are c^2 apart, because of $E = mc^2$. The object's mass is a small side effect about what happens in three-dimensional space, proportional to its energy, closely related to it, and also about resistance.

98. A rederivation of relativistic energy

You can say that, and to some the explanation might be neat, interesting, or even a beautiful one. But you've got to get to the mathematics - I'm sure you understand why.

The paper on rotation theory is called '*A Planck scale theory of time, with a fast early cosmological time rate, and rederivations for relativistic energy and time dilation*'. I'm not going to set out much of the mathematics in it, but a quick outline of one bit might be good, for those who'll understand, or get an idea of, the simple thing it does.

In physics, rederivations are important. A rederivation is about taking some mathematics that's on the map already, and showing how it can have come out of a particular picture. The picture might have been found decades after the mathematics, even a century. Sometimes we find the mathematics first, and the accompanying picture later, which is in a chapter in Book I, *'The cart and the horse'*.

Relativistic energy came out of Einstein's theory. As I've said, it's extremely unexplained: in relativity a velocity is nothing more than a viewpoint. And yet in experiments these frame-dependent energy changes are found to be real. So we need an explanation, and rotation theory explains it with the idea that energy is resistance to the flow.

So if you decide an object is moving in a certain way (ie. pick a frame), you're also choosing a direction for the circular flow. The object's particles are now 'spiralling' along a set of spinning tubes, which you've implied positions for. Looking at it closely, this led me to the existing mathematics of relativistic energy, giving support to the whole picture.

For a measure of how much a moving object's matter is resisting the flow as it travels, the best indicator is its rotation rate, r. That's one of two rotation rates that go in opposite directions: the background rotation, which is always at c, and the object's rotation rate (one of its two speed components), which is always below c.

When a particle is at rest, and not moving in ordinary space, these two rates are the same. You get standing waves - running to stand still - and that's why the two rotation speeds go in opposite directions.

But when it moves, you get extra resistance, and *the ratio between these two rotation speeds* quantifies it. They push in opposite directions, but they have the same orientation in space, which means they can be compared. It's not a fixed orientation, but their ratio still shows how much the object is 'pushing upstream', against the flow. The more it pushes upstream, the less the object has to resist the flow. Standing waves, standing still, don't have to resist the flow much. Pushing upstream helps them. But moving sideways, the object is pushing upstream less, so it has to resist the flow more.

I'll show how the mathematics of relativistic energy can be rederived out of that. If the object's velocity is v, its rotation rate r in m/sec., in the direction of the background flow, from speed components, ie. $v^2 + r^2 = c^2$, is

$$r = \sqrt{c^2 - v^2} .$$ (5)

The ratio between the two rotation rates I mentioned is c/r. That number is a measure of matter's resistance to the flow, and if that also measures energy, matter's rest energy E_0 by that gives the total energy E. That's $E_0 \times c/r = E$.

The ratio c/r can be expressed, from equation 5 above, as

$$c/\sqrt{c^2 - v^2} .$$ (6)

The rest energy E_0 is the rest mass m_0 by c^2 (the mass-energy equivalence) so multiplying $m_0 c^2$ by equation 6, you get rest energy x relativistic energy:

$$E = m_0 c^3 / \sqrt{c^2 - v^2} .$$ (7)

That's the total energy, including relativistic energy, as derived from rotation theory. In standard physics, the total energy is very often taken to be:

$$E = \gamma_{(v)} m_0 c^2$$ (8)

where

$$\gamma_{(v)} = 1/\sqrt{1 - (v/c)^2} .$$ (9)

So the result is

$$E = m_0 c^2 / \sqrt{1 - (v/c)^2} .$$ (10)

Equations 7 and 10, from RT and SR, turn out to be exactly the same, but just expressed in different ways. So this is an interpretation for relativistic energy - both for the concepts (what's called the observed phenomenology) and for the mathematics. It's straightforward to rederive relativistic mass in exactly the same way, via the definition of mass from rotation theory.

Time dilation can also be rederived from the basic picture, and in the paper I've done that. So key areas of special relativity come out of rotation theory, and hopefully this shows how it works, and what's going on.

99. Standing back

Standing back and looking at this whole set of explanations, it works as one single explanation. There's a lot of cross-corroboration: the different parts of it support each other in many places. Further connections are made if ideas from Books I and II are put in alongside these.

The definitions for energy, inertia and mass I've given show that only one of them is truly universal. As Richard Feynman pointed out, the rules for energy apply to everything without exception - light is included. And so is everything else: if something resists the flow, no-one minds what, where or why - it has energy, and that's it. But it's quantifiable, and that brings an enormous range of activity together under one umbrella.

The picture that emerges explains a lot, including one particular part of the mass-energy equivalence that's hard to explain. It's about what happens with the internal energy of a stationary particle. When matter is stationary, it just sits there, and there doesn't seem to be enough happening for us to explain how the mass-energy relationship is maintained. We've got right down to a single particle, there's not much to work with, but there's still this universal relationship between mass and energy - it's hard to build an explanation out of what little is there. But with this idea of matter as disturbances in a flow, rest mass and rest energy fit closely together, because there's hidden motion happening, and hidden resistance to that motion.

The next section is about the things that can be seen with hindsight, now the picture is in place. If the picture is right, a lot of jigsaw pieces click into place (and when I'm on the wrong track, nothing does). You may have noticed this clicking into place in this book, I hope so. That what a picture really should do. So if it's right, the hindsight section will say a lot, and hopefully shed light on some very old puzzles, as well as some new ones.

Section III. Hindsight

Part 23. Accelerated motion

100. Concepts and pictures

The explanations in the last section all come from a single picture. I'll make a wider point about pictures, as the last few chapters showed strong evidence for these ideas, and hopefully the next few will reinforce them.

If the picture I've shown is right, physics stayed stuck for a long time because a conceptual breakthrough was needed, but it was not geared up to making one. What do physicists spend most of their time looking at? It's not the raw unexplained clues. In many cases, it's existing mathematical theory. People are combing through areas of the jigsaw that seem to be in place already. So although they might of course find new ways forward from there, and build bridges out across the holes in the jigsaw, the picture is still unclear, and we need to work on it.

And physics courses are focussed on the mathematical side: students are not encouraged to ask 'why', or look for concepts. The implication is often that we know these things already. And what is usually given credit in physics has the same emphasis, in many places.

That's why, for instance, on every page actually about Simon Kochen on the internet, there's no mention of his greatest achievement: he's the discoverer of relational quantum mechanics. But others did the mathematics, and he's an excellent mathematician himself. So when they list what he's done in his career, his enormous breakthrough of 1979 *isn't even mentioned*. But it was arguably the main breakthrough in QM since Bell's, perhaps since 1926, and that kind of discovery is exactly what physics needs now.

Sometimes it's like a house full of electricians, trying to solve a problem that turns out to be a plumbing problem. If a plumber comes along and solves it, they don't all accept it, and some of them say 'he's not even an electrician!' (In fact he may have studied both wiring and plumbing, rather as some study mathematical physics and philosophy.) I tell you this because the future may bring some similar scenario.

In a more general way, people are often against anything that *seems* to make their own years of study less valuable. Although outsiders from the academic system have contributed enormously to physics - Leibniz, Faraday, Gallileo, and in biology Darwin - and were better equipped to think outside the box by living outside it, this aspect of progress is often ignored.

Einstein had no academic affiliation when he published special relativity, and was an outsider at the time. He was an independent researcher, like myself and others, during his 'annus mirabilis' in which he published several papers, including on special relativity. The famous mathematician David Hilbert once told a conference of mathematicians: *'Do you know why Einstein said the most original and profound things about space and time that have been said in our generation? Because he had learned nothing about all the philosophy and mathematics of space and time.'*

So outsiders can be important. But they're still discriminated against, even in the 21st century, where we value non-discrimination highly - I'll explain about that later. We can't afford to throw out the work of outsiders, physics rests on it. But to some it appears to diminish their own learning path. In fact, we all make a choice. The deeper reality is, each to their own - different learning paths are good for different people.

Here's what may be the same thing in another area. Shakespeare wrote his plays with very little formal education, but he did a lot of reading. Some may have seen this as a threat to their own education, because if he did that with no education, it might seem to make their own path less worthwhile. So they decided there must have been a well-educated aristocrat, who wrote all the plays anonymously, supporting literary education as he went.

They found no direct evidence, and the main suspect they had for writing the plays died before some very good ones were written. But they still somehow managed to conclude that the Stratford man who went down to London and started a theatre named *The Globe Theatre*, was a different person from the person who wrote 'All the world's a stage' - characteristic as the speech in *As you like it* is, along with one on similar lines in *Macbeth*. It's true the phrase existed already, but Shakespeare adopted it closely, in the name and motto of his theatre, and in his writing. So the theatre and the plays look very much like the work of one man.

Now there are many arguments on this, and I don't want to dismiss the other ones too easily. But what I'm saying is that this kind of motivator - *if present* - although not a conscious one, can be a strong force. What may pass for open minded reasoning can have an end goal underneath it. So having pointed out this tendency to try to defend years of study, I hope we'll watch out for the same thing in physics, where those who take a less conventional learning route - although outsiders have contributed enormously in the past - may be criticised or have their work blocked, for similar reasons.

And this is part of a wider pattern. We feel comfortable in certain areas, like the mathematics, and the emphasis reflects this. The conceptual side can be less easy - it's often hard to explain things. But that's where we might make real progress.

Attempts to downgrade the conceptual side during the 20th century showed this clearly. 'Shut up and calculate' became a catchphrase. The areas of any puzzle that look uncomfortable go under the carpet, though they're the very areas we need to be looking at. But good physicists, by contrast, tend to give lower priority to those kind of things. What they want, as always, is simply to know what's out there.

101. Newton's bucket

So now we get to hindsight, and we can try it out on a few existing ideas. The first thing is to take a fresh look at a small puzzle about inertia. Well, it looks like a small puzzle - on the face of it the issue is almost laughable. But oddly enough, it's crucial to our view of how the universe works, and people have been arguing about it for several hundred years.

Newton pointed out that if a bucket of water is rotating, the water will move up the sides of the bucket. He asked the question, how does the water know it's rotating in order to do that? What is it rotating in relation to?

This issue impacts on some very deep questions. It's about whether motion is relative or absolute, so it affects the question of whether space has some existence without matter. Galileo had already pointed out that some motion is relative. But if *all* motion is relative, then the water shouldn't know the difference between a rotating bucket and a non-rotating one. But the water responds to something - it moves up the sides, and travels around like that (large-scale rotation). And as Newton pointed out at the time, if the bucket is stopped from spinning, the water will carry on rotating for a while, still with a concave surface.

Newton got his hands on a bucket, carried out the experiment, and used it to argue for absolute motion in relation to the aether. Though we now disagree with some of what he thought, Newton had found a brilliant way to put his finger on a wider question, finding a distillation of the problem. Because the water changes shape, you have something tangible to be explained.

There are two kinds of motion, and they seem different. Motion in a straight line at a constant speed is relative, and from some other viewpoint or frame, it might not exist. That kind of motion can be made to disappear, simply by looking at the scene from another viewpoint. So there's an interchangeability between being at rest, and moving in a straight line at a constant speed.

But *changing* motion, like in the bucket experiment - accelerated motion - doesn't disappear so easily. Whenever matter travels around a curve, inertial forces pull it in specific directions. This kind of motion behaves in some ways as if it's not just relative, there seems to be an element of absolute motion involved as well.

The unanswered question, with accelerated motion, is how does matter get its bearings - how does it know which directions to be pulled in by inertial forces? This might seem obvious, but it has been a problem both before and after relativity. As was mentioned, the best solution we have now is Mach's principle, in which matter gets its bearings from the gravity of all the distant galaxies, or the entire gravitational field of the universe.

Although Newton believed the motion of the water in the bucket is absolute, others didn't think so. Huygens, Leibniz, Berkeley and Mach argued that it's only relative, and that the water responds because of its motion in relation to the other masses in the universe. When relativity came along, and people were breaking free from the idea of an aether, many thought that the way forward was for all motion to be entirely relative from then on.

This seemed to work well in some ways, but some liked it more than others. Einstein and Mach each thought highly of the other's work early on, but both changed their minds later, as some incompatibilities became more apparent. Various things are problematic about Mach's principle nowadays, and it has been hard to define, and hard to forge links between it and other areas of physics. It's an extra bit that was tacked onto the main body of 20th century physics, that doesn't fit with it, and raises difficult questions.

As for the problem of Newton's bucket, when general relativity arrived it had the potential to shed some light on it. The mathematics was difficult, but by the 1960s a complicated partial answer had been put together. It suggested that whether the universe was rotating (at a large scale) around the bucket, or whether the bucket itself was rotating, the water would go up the sides anyway. The trouble with this is not just that it sounds like something out of Monty Python - it also doesn't remove all the problems.

To some, this question reveals a deep crack in our picture. The issue sounds simple, but no reliable answer has been found. Some aspects of inertia seem very local - one problems with taking it as a long-range effect is a 'speed of gravity' issue. Does information from the distant galaxies arrive instantly, or at lightspeed? Neither answer is satisfactory - each raises further questions, nothing definite emerges. And explaining inertia via the far galaxies is failing, as inertia is extremely closely linked to mass, and the Higgs field shows that some particles respond to something locally to get their masses.

The background rotation solves both problems. It solves the bucket problem in the way it explains how particles get their individual masses. Both can be explained if matter responds to space locally, but to the background rotation rather than the Higgs field - which certainly exists in some form, but which at a deeper level may be an aspect of the background rotation.

The problem of how matter gets its bearings can be explained because space provides local 'reference points' for matter. Space has hidden motion going on in it, which provides directional information, even if it's relative. But some local directions are not relative: there are things that fix the landscape, such as the gravity field and the 'quantum framework'. So when matter changes its motion, and inertial forces arise, it knows which direction to be pulled in because it can tell what's happening from the background rotation.

This would mean the water in Newton's bucket didn't get its bearings from the far flung galaxies, as many people thought for a long time after Newton pointed it out. It got its bearings from the background rotation locally, and by 'locally', I mean right there inside the bucket.

If this is right, Newton couldn't have got there from what was known at the time. But he was pushing at vital questions, and his intuition about what was happening, according to this picture, was nearer the truth than Mach's view. Newton didn't think he was dealing with a long-range effect, and his view of inertia, with the Higgs boson showing us in 2012 that mass is at a small scale, now looks better than it did over the next few centuries.

102. Patterns in the waterfall

Before setting out the explanation for inertia, I mentioned that according to RT, every time your car goes around a curve, what you feel is the background rotation.

Now I can explain that further: when a car goes around a curve, the matter in it tries to keep going straight on. The reason is that the cylinders along which the matter has been 'spiralling' make straight lines, in the same sort of way as the road did before the curve. Those parallel straight lines are part of the structure of the dimensions, and they're what causes inertial motion always to happen in a straight line.

This straight line motion has never been explained, but it's a basic part of the universe. In this explanation, the world has something like a geometrical grid that underpins it. It's a framework, and it tells matter how it can move. But it's not only matter's position that is relative - the grid's position is as well, so that's why things get counterintuitive.

And going around a curve, with accelerated motion, the 'spiralling' rotational motion has to adjust itself, and change direction. For that to happen, the grid itself has to swing around. When inertial forces are felt, that readjustment is part of what's really happening.

Feeling the background rotation is like the forces that are felt when holding a spinning gyroscope in one's hand, and moving its orientation. The gyroscope tries to stay in one position, and gently resists attempts to shift its alignment. In fact, all matter does the same thing, but less. If you hold a brick, and then shift its position around, it also resists changes to its alignment a little. In this picture, instead of being like a large gyroscope, the brick is actually made of trillions of tiny gyroscope-like objects, all pointing in the same direction, all trying together to keep the brick pointing in a particular direction.

Newton's view of inertia was not incompatible with these ideas. He thought something inside the brick was causing a resistance, and as I said, he implied something else must be going on. That fits with RT, but in this view the cause is both internal and external, instead of the stark choice between the two, which people believed for centuries had to be made.

A loose analogy for this view of inertia involves a waterfall. I've talked about the patterns seen in a stream, or in the water from a tap, that are constantly recreated, and I've said that matter's duration through time is like that. The same happens with a waterfall. Thinking in that kind of way, and imagining the waterfall with shapes in its flow, an object that seems solid and enduring - a rock, a hotel, a bear - would be a pattern in the waterfall.

But it's an odd kind of waterfall, because patterns can move around in it. The flow is a circular one, and the patterns are standing waves, so it can retain patterns in a much better way than a stream or a waterfall can.

Now to get a picture of accelerated motion, here's a loose analogy. Imagine moving your hand through a waterfall. Moving it in a straight line, you might be less aware of the water. But move it around a curve, or change speed, and it becomes more noticeable. The point is, *changes* make the flow more apparent. This is not a close analogy at all, because either way you can feel the water on your hand.

But with matter in the universe, motion in a straight line at a constant speed is very like standing still - the dimensions' positions are interchangeable, with a basic symmetry between the axes, so it doesn't reveal the rotation. There's enough ambiguity there to remove the motion simply by deciding the object is in its rest frame, which swings the axes around to a new position. So that kind of motion, inertial motion, only exists at all if you specify exactly how an object is moving.

But the other kind of motion, accelerated motion, isn't removed in this way. The waterfall analogy is limited, but it still illustrates loosely why in the real world, if you start moving around curves or changing speed - accelerating, that is - you feel inertial forces. It's because that kind of motion gives away the hidden background motion that is going on behind everything else, and through which everything that moves moves.

But more accurately, the difference between the two kinds of motion is that one kind - inertial motion - has the dimensions keeping their positions. Once they're decided on, the object can move, but the dimensions don't. But the other kind - accelerated motion - has the dimensions changing positions, and swinging around, even as the object moves.

103. Why do clocks get out of sync?

If you look back across the picture so far, a landscape should emerge that has self-consistency. It's unfamiliar, but it hangs together well, and more jigsaw pieces will hopefully click into place with hindsight - they did for me anyway. The concepts should cross-corroborate and reinforce each other.

So let's take a look at one or two more puzzles in physics, hopefully assisted by hindsight. Special relativity is a natural starting point. People have been in confused discussions on it ever since it first appeared.

One aspect of special relativity is an observer-related effect according to this picture. For a loose analogy, it's like standing in a field looking at the clouds, as I was the other day. When the clouds run parallel (as they say they do in Leipzig), it can seem that this is a real pattern.

But wait a minute: the parallel lines are always at right angles to a line from the observer's eye to the horizon. You can turn around slowly and look in any direction - the clouds always make parallel lines at right angles to your line of sight. So the effect is observer-related, and it's about a foreshortening effect, compressing the shapes of the clouds along the line of sight in all directions, making them look elongated sideways.

Special relativity is like this kind of optical illusion. But it also has elements of reality in it, it's a mixture. And that has been one of the hardest things to get to grips with, and to explain.

In our present view of special relativity, the only aspect of the observer that's thought to contribute is their state of motion. But in RT you also sometimes have to take into account what the observer is made of - matter. And that exists on the dimensions, and isn't what it seems.

In RT, as in special relativity, two observers passing each other will each see the other in slow motion, as if their viewpoints are inconsistent (Chapter 4). But although in symmetrical situations it's exactly as if there's some kind of illusion, in asymmetrical situations age differences will remain.

The idea of the constant replacement of matter allows this mixture of illusion and reality. Both observers are being replaced by other identical observers. This is an odd way of putting it, but it happens at a very small scale, and very rapidly. In our familiar large-scale world people and objects have continuity, like the people and objects in a film. But if we could see to the Planck scale, just like people and objects in a film, we'd see that they're being replaced. So when two observers get out of sync with each other, one of them has been getting replaced faster than the other.

Say you get two film projectors, and point them at areas of the same screen, and run bits of the film at different speeds. Things would get out of sync, and the only way to understand this would be to understand the mechanism that is sitting there underneath what you're seeing.

It might be easier to imagine two clocks, which we often do when trying to understand these things. The clocks get out of sync because they're moving differently, or at different heights in a gravity field - we know that much from experiments. According to rotation theory, what happens is that both clocks are steadily being replaced, but at a very small scale. And if it's happening at two different rates - for instance if one clock is moving, and helical motion at that scale is slowing the replacement process down - then it will tick slower than the other.

104. Hidden motion in a static picture

Standing further back, the difficulty getting to grips with time, according to this picture, is mainly because we've been mixing two levels. These levels are real (and not the same as the two contradictory levels of time in Chapter 31 - one frozen, the other with motion).

To understand time in rotation theory, there's a need to separate two levels. Newton's unchanging time, the deeper level, has to be separated off from Einstein's variable time, the superficial level. But the word 'time' can be kept for the superficial level, which is its normal use anyway, because we hardly ever need to refer to the deeper level.

This mixing of levels explains most of the confusion about time, and many of the contradictions. It arose because in the early 20th century, we interpreted special relativity wrongly in relation to time - nothing wrong with that, that's not a crime. What is a crime (metaphorically speaking), is that we failed to *leave room* for those ideas to be wrong.

Another point that comes with the benefit of hindsight, if one believes RT is right, is a general one about lateral jumps. For whatever reason, one kind of lateral jump is very often unexpected - the kind with the addition of unseen motion. Realising there's hidden motion happening in an apparently static scene is not one of our strongest points.

The idea that the universe was expanding was missed until the evidence for it was found. Few thought of it - even Einstein didn't, who was far from being one to accept unquestioningly the current preconceptions. With hindsight it was very likely that the universe was expanding or contracting - if not, gravity would pull it together. The way they imagined it, it would start contracting. But the picture of a static, unchanging universe was deeply ingrained at the time, and the idea that everything might be moving was unexpected. And it seems generally that somehow, adding invisible motion into a static picture doesn't come naturally to us.

Perhaps it goes back to when our ancestors were like those meercats we see on TV (no, not those ones), always on the lookout for sudden movement. It would often mean problems, so it's not what we hope to see. But in physics it's the opposite - unexpected motion can be a solution to problems.

Refractive medium gravity goes back particularly to Eddington. People knew the medium had to get thinner as you go outwards. But I'm not heard of an emitted medium that dissipates, so the field is always the same shape, but is constantly moving outwards, except in my own work.

And in foundational physics, circular motion at the Planck scale shouldn't be totally surprising. At every other scale we find the universe is in motion. And a large part of that motion is rotational motion, which happens easily, given half a chance. And because when matter changes shape, it naturally settles into spheres or disk shapes - and dimensions now look like they probably do the same - it very often is given half a chance.

Part 24. Two puzzles solved with one idea

105. The gulf across scales

One of the most important questions is all physics is simply 'What's it like at the Planck scale?' And although we find a list of things that hint at some kind of deep regularity sitting right there underneath the world, current physics has not been able to describe the Planck scale in that way.

Present ideas of the Planck scale vary, but picture is usually disordered. We have even less of a picture than a few years ago, as supersymmetry has been shown to be wrong, which means the picture from string theory is unreliable, to put it mildly. Our theories need to come together there, but they've been having trouble doing that. At larger scales, unifying gravity with other physics is our worst problem, but at that scale, it's even more of one.

Going by standard theory, the world down there is chaotic. Quantum effects dominate, energy decays into particles and annihilates, virtual particles and perhaps very small black holes are constantly created. The fabric of what we call spacetime deteriorates into an uneven foam, structure rapidly appears and disappears. The result might be a sea of virtual black holes, or a foam of bubbles - as the line from an old song in a musical goes: '....*disorder, chaos, disintegration! In short, you have a ghastly mess*'.

This is in stark contrast with the highly ordered picture from RT. If the chaotic picture is accurate, it's odd that things are so ordered at larger scales. Small-scale chaos might be expected to lead to general chaos. But although things break down at the Planck scale in standard physics, this might just mean that our physics breaks down there, rather than the world.

One reason disorder is expected there is about the current mechanism for gravity, space curvature. At the Planck scale it creates gravity strong enough to curve space around on itself into bubbles, which bubble around as part of a general disorder. But if a there's a different mechanism for gravity - PSG or one of the other new theories that are being taken more seriously nowadays - it might solve the problem. An alternative gravity theory could make a more ordered world at the Planck scale possible, which would allow the large-scale world to emerge from it. So a change to our view of gravity might help us to connect up the picture properly.

If there's anything beyond the Planck scale (there probably is), string theory suggests what we call space and time don't exist. They seem to get started at the Planck scale. Brian Greene says: *'...the familiar notions of space and time do not extend into the sub-Planckian realm, which suggests that space and time as we currently understand them may be mere approximations to more fundamental concepts that still await our discovery.'*

Most physics is completely stumped if it tries to explain this, but it fits the RT picture well. Space and time appear at the Planck scale (or some equivalent scale), because that's where the rotation happens. The background rotation creates what we call time, because of the constant replacement of matter, and it also creates space - as we know it.

It's a setup that generates something like spacetime, as soon as you add the background rotation in, and the accompanying picture of matter. The setup gets amplified to a large scale, and spacetime is generated out of what Paul Davies called 'pre-geometry', when looking at the surrounding possibilities. He didn't say anything about what it would be, but he thought there might be something that initially generates the geometry at a small scale.

Along with space and time as we know them, the rotation also creates all the regularity and symmetry in the behaviour of matter - which hints, in so many areas of physics, that there's not just a chaotic mess, but deeper underlying machinery down there.

What do we know about this underlying machinery? It creates various things that we haven't been able to explain so far, like energy and inertia. These go by simple mathematical rules. It's also extremely precise: the near-proof set out in Appendix D, surprisingly enough, proves beyond much doubt that PSG is right, or largely right. And that means matter gets refracted at that kind of scale, along helical paths. For refraction to work as it does, a light beam must hit a boundary with one side of the beam arriving first - this turns it through an angle. So these incredibly small-scale waves have 'edges' - and that means the machinery down there is already looking very precise.

106. A new angle on an old argument

This overall picture of space and time should be put into context. It brings a genuinely new way of seeing things into a very old argument. It started with the question of whether space is something or nothing, and whether it exists independently of matter.

The argument has roots that go back to the Greek philosophers, and Zeno in the 5th century BC, at which point they started arguing it out. The argument, you might say, was about whether space is the final frontier (and no doubt some accused others of being 'illogical'), to put it in modern terms.

At first the two views were simple enough. In the absolutist view, space was something real that existed, because objects could move or have positions in relation to it. But in the relational view space doesn't exist, and the motion or positioning of objects was purely relative to other objects.

The argument had got more detailed and specific by the time Newton and Leibniz got hold of it, and argued it out between them, with Newton taking the absolutist view, and Leibniz taking the relational one.

But we've always had trouble with this basic question of absolute or relative motion. The two early ways of seeing it were extremes, but as we've learned more about the world, instead of being able to dispense with one or the other, increasingly subtle blends of the two views have been formulated. It seems that either on its own is not enough, and that we can't do without at least an element of both.

The argument hinged on the question of whether there's a 'preferred frame' in the universe - a frame more significant than others, against which motion can be defined. Clearly if a space has edges, then there's a preferred frame in which objects move in relation to the edges. When we look at a pool table, we use the edges to judge the balls' positions, and their motion. But without any definite news from the edge of the universe, people have had to look at the difficult question of whether space has enough existence in its own right to define matter's motion.

And no clear solutions have appeared. As we learned more, the vital issues became more and more complicated, and some have remained unresolved - including recent questions. Special relativity had no preferred frame as one of its main premises, so the relationships between objects are all-important, as if that's all there is, and space has no real existence.

But general relativity, by contrast, suggests space does have an existence - it can bend and stretch, and gets distorted out of shape by matter. So in terms of that old argument, relativity seems to pull in both directions. But a lot of things have been pulling in both directions for a long time. It seems that both concepts will have to be involved somehow, if we're to get to the real, true picture. But to add in both absolute and relative motion in the right way, a lateral jump might be needed.

Well here's one, and I hope it makes sense. In the picture of the world from rotation theory, space definitely exists, with a definite structure, and matter is irregularities in it. But the rotational motion smooths it back into regularity again, so matter doesn't behave like irregularities in anything. Instead it gives the impression of being completely separate from space. But because matter is really part of space, its motion in space, and the way it behaves, isn't what we expect. This allows a new kind of solution.

And in this solution, both light and matter have simple motion in relation to a transmitting medium, as (apart from in a gravity field) they always move at the same speed in relation to it. So far so good, but the transmitting medium is in rapid motion. That's a lateral idea in itself, but on top of that here's the really unexpected twist - *the medium only has a relative positioning in space*. So motion is normal, but in relation to a medium that is not.

So the weirdness is far more in the transmitting medium (Chapter 25), than in what moves through it. This approach is new - it's a particular combination of what in a sense is related to absolute and relative motion, but it's a lateral jump that hasn't been suggested so far.

107. Apples to oranges: a funny thing about frames

On Earth, it's easy to tell if the train or the platform is moving. It's the train. What picks this out (apart from common sense, which is sometimes wrong), is the fact that the train accelerates away from the station and decelerates into the next one.

But although the truer answer is that the train is what moves, there's still an inbuilt bias towards one particular frame. We relate to it a lot: in it, the Earth is fixed, and everything else moves on it, and across it. In the past, that frame has affected other issues: when some religious people gave Galileo problems about his view that the Earth moves around the Sun, in one way of looking at it, they took offence at a reference frame.

And although when the train pulls away the station seems to stay completely still, it doesn't. Liverpool Street Station is moving around the Earth, around the Sun, around the galaxy - and onwards and outwards, following a looping, curving, crazy trajectory, which is unexpected for a London mainline station, and cannot be found in the helpful timetables there, or by asking a station guard or a policeman.

In the universe everything is moving, so you need to assume something isn't. Then a lot more can be worked out, via a system of frames. We made some assumptions to do that, and they seemed very reasonable. But one thing was wrong. I'm going to show what was summarised in Chapter 43: large areas of the problems we've had in physics in the last hundred years arose, amazingly enough, from a single false assumption about frames.

Reference frames go back at least to Galileo, who first put relative motion on the map. He saw that if two boats pass each other, either could be moving. You can compare the boats' motion with the water, but in space you can't do that. Ultimately it seems there are just objects in space, and they move only in relation to each other.

In the picture from RT, an object's rest frame is a viewpoint with its waves on circular paths. In another frame it's moving, and they're on helical paths. The waves are wrapped around a cylinder, and how the cylinder is positioned is a viewpoint.

This might seem the same as the normal idea of frames, but just with a new picture attached. But it's different. You can relate frames in one way, but not in another. The dimensions can't be set at two orientations at once. You can transform between them, swinging the axes around to another position. This doesn't cause a problem, if you don't take the two viewpoints to be true at once. But if you do, you'll be comparing apples to oranges.

This explains something we know about, but very rarely talk about. In special relativity, frames can't be linked *at all* via their 'now' moments. They seem to be utterly incompatible. I've done this calculation, but it's known.

Take two lorries in the desert, and load a very accurate clock onto each. Send them off on different roads at different angles, and at different speeds. Then calculate what time it is *now* on one lorry, from the viewpoint of the other, using basic special relativity. It's not hard to get an exact time. Then do the reverse calculation, using that 'now' moment as a starting point. You find the reverse calculation doesn't take you back to the 'now' moment you originally started from. They're not linked in the way we'd assume. You can even have a telescope on one lorry, look at the clock on the other, and allow for light travel times - the result still won't be consistent.

So 'now' moments in different frames *don't match up*. This never bothered people much, as now moments are not meant to exist. But we use them all the time, exactly as if they did exist, in many vital areas of physics. And you'd expect them to have *some* relationship, even if going by weird internal logic. Simultaneity in special relativity has that in many places - it usually has some sort of self-consistency. But here it falls apart.

According to RT, the reason frames can be totally incompatible is because in a different frame, the universe is being seen in an entirely different way, with the dimensions in different positions.

But some inertial frames have a simple interchangeability. In RT, two objects can be related in either of two ways: with one in its rest frame and the other moving, or vice versa. In this kind of situation, swinging the dimensions from one position around to another, is enough to make motion disappear. This is usually exactly as frames are in standard physics.

But then we finally hit a real, major difference. It becomes clear that unlike in standard physics, if there are more than two objects, more than one frame will be needed. This is where the views diverge in a major way.

Imagine an observer watching two objects. With the observer, that's three. The two are moving in different directions, on straight line paths, at different speeds. The observer decides she isn't moving. From that viewpoint, object A is moving, say, upwards, to the left, and receding - while object B is travelling diagonally downwards and to the right.

In standard physics this can all be in a single frame. That seems reasonable, but in rotation theory it's two. The motion of object A needs the dimensions in one position, but object B needs them in another. So it's like when people want to watch different channels on the same TV: objects A and B both want the dimensions, but only one can have them at a time. So each object has to be related to the observer separately.

And in standard physics, we find we do exactly that, in another way. We only relate the motion of one object at a time to the motion of light. It's a lateral jump to think of it that way, and no-one did at the time this was discovered, around 120 years ago.

Everyone assumed, naturally enough, that all objects move in the way they do in relation to light *all at the same time*. It would have been a very strange idea to say that we have to relate them to the motion of light separately, as if they were not all part of a single picture. But they're not: if the dimensions have to be put in different positions for different bits of the picture, then the picture comes apart into many separate areas. We'll return to the motion of light and matter in the chapter after next.

So according to RT, frames are different. But oddly enough, the differences usually do nothing. They 'come out in the wash': special relativity has ways to compensate for them, and transform (translate) between viewpoints. And when you get right down to it, special relativity itself can be seen as a system of transformations between positionings for the dimensions.

108. The hidden flaw in the three steps

So very often - more or less always in fact - it turns out that two incompatible scenarios are not actually being taken together. What usually happens is that instead they land near to each other in the reasoning.

But step two in the three steps that led to block time is a massive exception. What I've called the three steps (Chapter 23), is a simplified version of what's known as the Rietdijk-Putnam argument. Step two means doing something that according to rotation theory, you just can't do.

It involves relating three objects, in a single frame. And according to rotation theory that entire argument depends on an approach that simply isn't valid: two incompatible frames are being taken as a single frame. This is at the very heart of why we got a false picture of time.

The argument that led to block time, from which so much confusion arose, in a simple form that Roger Penrose has described, involves two people passing in the street. It's about the rules we've assumed to be true on simultaneity in relativity, and it leads to the three steps.

A distant event happens, say in the next galaxy. One person is moving away from it, the other towards it, as they pass. Penrose points out that according to the Minkowskian geometry, the event can be in the past for one person, but in the future for the other. Being in the past, that event is inevitable - but if the same event is also in the future from another point of view, that makes a future event inevitable. This gave us an intractable problem, which we've been trying to deal with for a hundred years.

But to imagine that situation and draw that conclusion, you're relating three objects, not two, that are all moving differently. Two observers, and a distant event. That's fine in standard physics: a frame can have all kinds of objects moving differently - as many as you want to put in it. But in rotation theory, three involves superimposing two separate views on the universe. And that's apples to oranges. So you get no reliable information out of that, and it's a false conclusion that an event can be in the past for one observer, but in the future for the other.

In my version of things, you have to take that one frame with three objects, and break it up into two frames, each with two objects. The two observers can then each be related to the distant event, but the two relationships with it must be kept separate. You're not allowed to draw conclusions about both of them taken together. *But what led to block time did.*

And that single point turns out to make a surprisingly big difference: it's like the first domino in a long line of them. Our whole view of time depends on, and hinges on, that one idea. If step two goes, then block time goes with it. Instead you have relationships between things that certainly connect up, but not in the way that has been assumed.

And with hindsight, matter's motion can be taken to need relating *separately* to the motion of light. It's one way of looking at it, but it solves two problems that have been hard to solve for a long time. Perhaps what I said in Chapter 25, about how special relativity leaves room for the underlying assumptions themselves to be different, now makes better sense.

If this is right, it slays the dragon of block time, because all our reasoning that led to block time depended on relating frames in that way. And all of our reasoning that led on from there, including the idea that time doesn't exist, and that the motion through time we observe is some illusion, or emergent, also depended on that false conclusion.

If this view is right it opens the floodgates, removing the restrictive logic that seemed to make block time inevitable. It frees up the possibility of motion through time, and an undecided future - and that means a future compatible with the randomness of quantum events.

And that liberating mechanism also makes it possible to remove a long list of contradictions about time, all of which arose from a false chain of reasoning. So it allows the landscape to go through real change, and for our thinking to go onwards and outwards from there, and hopefully towards something that works a lot better.

109. The motion of light and matter

As I said, this point about frames also solves another puzzle. It's a major one. Not everyone thinks it has an explanation - our present explanation is 'that's just the way it is'. Few, if any, other explanations have been suggested. But if you find an explanation that works, it makes it more likely that it does have an explanation. The problem is actually what started all these questions off in the first place, more than a century ago.

It's well known that however an object is moving, light moves at c in relation to it. You can chase a light beam, or run away from one, but the light beam still moves at the same speed in relation to you.

But that can be misleading: if you talk about running in some direction, it has to be in relation to something. Best to start standing still. In any object's rest frame (when it's not moving), light moves at c. That's not too surprising. For someone onboard the object, its rest frame is the natural frame from which to measure the speed of anything going past it. Stand on a chunk of rock out in space - all you do is take your chunk of rock to be still, and you find light moves past it at the speed of light.

So far so good. In RT, it's exactly the same, light moves at c in an object's rest frame. Here's what it looks like: light travels at c along parallel cylinders. The object is at rest, so its matter makes circles. To relate the object to the light, both must be on the cylinders, so they're placed into the picture, and angled correctly. The planes of the circles are at right angles to the light's motion, so light moves at c in the object's rest frame.

This adjustability of the dimensions is unexpected, but they've always been adjustable. It's just that we didn't see matter as waves in their fabric. So this version of things, with light and one object, produces nothing that might be referred to as 'totally weird'.

But it starts to head that way when you add another object into the picture, moving differently from the first. It then seems amazing that the light moves at c in relation to both, as the two objects are also moving in relation to each other. In standard physics, this is just a fact that has to be accepted.

But in rotation theory there's an explanation. The motion of light can only be related to that of matter via an object's rest frame, and *only one object at a time*. So with two differently moving objects, it's true that light moves at c in relation to the rest frames of both. But these two relationships are separate, because each relationship involves a positioning for the dimensions, and the two positionings are incompatible. This breaks the picture up into more than one element - if it's right, it means the confusion arises from superimposing separate frames, which are incompatible viewpoints.

I started to think about this one day when looking across a large, wide field in Surrey. There was a tractor moving in one place, a plane way off to the side, seagulls moving in different directions in another part of the picture. There were also those short cylindrical rolls of hay in the field, positioned at various angles. According to the theory I was thinking about, our view on the world subdivides into separate areas, and they're not connected up in the way we assume. We assume what we see has a unity to it, and is connected up in a certain way. But it isn't, it's connected up in a different way.

In this approach, a scene with differently moving objects has many different positionings for the dimensions, and it goes right down to the motion of dust particles, and beyond. This idea of dividing the picture into smaller sections is hard to understand - it's all very well to say that we can't take these things together, but we naturally do take many things together. It brings us to the question of what's really out there.

And what's really out there is a world with various viewpoints on it, which we call reference frames. If we use a different system of frames from the one we use now, most of what gets changed is the way we look at things.

So going back to the view across the field, say one of the seagulls decides to fly in a particular direction. A new viewpoint will immediately exist, with the ground still, the bird's matter on helical paths, and the dimensions in certain positions. And if someone on the ground measured the seagull's speed, they could calculate its helical path angle in relation to them.

To get to grips better with light moving in relation to matter, I'll talk about an experiment that was actually done. When the technology arrived, one of the many experiments on light and matter happened at CERN in 1964.

Neutral pion particles were created, and sent off at $0.99975c$, a little below lightspeed. They were then made to emit gamma rays (which we often call 'light'), while moving. So matter moving very fast emitted light, which moved off ahead of it, in the same direction. The light was measured to be moving at c in the lab frame, as it always does, and not c plus the emission velocity, or $1.99975c$.

This is standard stuff in special relativity. In the lab frame, the light seems to move slightly faster than the pions: they're measured to be moving at c and $0.99975c$ respectively. You'd expect the light to move away from the pions at $0.00025c$, which is rather slow. But in the pions' rest frame, the light moves away from them at c.

This doesn't make intuitive sense. Many believe it isn't meant to: the world just *is* that way. But with the clues we have about space being a transmitting medium, and many other clues about space having a structure to it, it seems possible, to put it mildly, that there's more to be found out.

Anyway, here's one picture from rotation theory, of what was really going on in that lab in 1964. The pions, travelling at $0.99975c$ in the lab frame, are on very stretched out helical paths. The angle on the helical paths, arcsine (v/c), is 88.7188 degrees. They're not far from 90°, they nearly make straight lines along the length of their cylinders, although they can't quite do that.

That's the starting scenario: the pions travelling on very stretched-out helical paths. The gamma rays, once they're emitted forward, travel away in straight lines at c, along the length of the cylinder.

Both the light and the pions are waves moving at identical speeds through a medium, both at its basic transmission speed, and in very similar directions. So it's unsurprising that the light doesn't take on the emission speed as well as its own speed. The slight difference to their *measured* speed (their 'visible speed'), arises only because the pions occasionally have to travel around the cylinder, being made of matter. Their path is set at a slightly different angle, which affects their visible speed. So you have two sets of ripples, matter and light, moving at the same speed, but at slightly different angles.

Strictly speaking, you can only relate the light to the pions in their rest frame, so mathematically, it starts from a different picture. But hopefully this gives an idea of what's going on.

All of this goes with a change to the rules about frames, and it's worth saying that the adjustment removes two central problems in physics. They're not at all easy to remove, and each solution supports the other, because in rotation theory they use the same escape route.

One is the contradictions that come with block time, which disappear if the Rietdijk-Putnam argument is false. The other is the major difficulty explaining light and matter's motion, which is removed in exactly the same way. So this difference to frames, which has solid reasoning behind it - about matter and the dimensions - starts to look like an effective tool for change. But the odd thing is, in most situations, the adjustment hardly affects anything. In general the differences make no difference. If they did, it would have been noticed decades ago.

But they do get crossed wires with Minkowski's geometry in one particular place. It's unfortunate, according to my view, that this area led to our whole view of time. It just happened to lead to a small group of assumptions, which then led to a larger group of assumptions. But what was really far worse was, it then led to an even larger group of assumptions.

And as in Chapter 24, a single false idea can be like a spanner in the works, causing problems in many areas. If the view here is right, it will be possible to leave these problems behind, which have caused confusion in many places, and start seeing the world with fresh eyes.

Part 25. Summing up

110. The different parts of the theory

You've now got the underlying picture, which is a backdrop that goes behind the whole theory. I'll make it clear what the different parts of it are, to avoid confusion.

The whole theory is rotation theory, RT. There were three parts of that, each of which led to a paper and a book. They were done in reverse order, having been found in the order: time, gravity, quantum mechanics. The books were published in the order: quantum mechanics, gravity, time. I'll tell you why in a minute, it was partly just how it happened.

The books were called *The Unsolved Puzzle: Interactions not measurements*, *Gravity: what we still don't know*, and *Time: Is the future already decided*? The three sub-theories that make up RT also came out in papers published in reverse order: DQM, PSG, PST, dimensional quantum mechanics, Planck scale gravity, Planck scale time.

The theory in this book was kept until last, Planck scale time theory, most of which is an interpretation for special relativity. PST is one of three pictures, but it's the ancestor of the other two, having started off first, in the mid '90s. It's the central part of rotation theory in terms of the picture, but PSG is in terms of the mathematics, which came second. Third came DQM, which is in Book I, as it made sense to start with it.

One reason was that a documentary was being made about DQM, because a friend and documentary maker, Nigel Lesmoir-Gordon, became interested in it. He suggested starting with a book on quantum theory, to accompany the documentary he was making, having sent my paper to Carlo Rovelli and Neil Turok. They both agreed to discuss the theory in filmed conversations - the result was 'The Interactions Avenue', which is online.

But a more general reason was that of the three puzzles I've tried to tackle, quantum mechanics is where the need for a solution was most obvious. So I didn't have to start by convincing people there's a puzzle at all. In quantum theory, the need for a solution is not half swept under the carpet, as it is with some other puzzles - everyone knows about it. In fact, with time, nowadays many know about that puzzle as well.

But few were expecting any solution to appear, and this applies to both time and quantum mechanics. People would wax lyrical about time. Books were written that seem to say a lot, in a 'poetic' way, while actually saying almost nothing: offering no solutions, no new physics, and avoiding the real clues, as they give away how little we know. Physicists felt safe doing that. With time, no-one thought new physics would come along in their lifetime and change the landscape in a major way.

111. The common potoo bird

But it might. I want to point out a similarity between the two main puzzles of the last century: time and quantum mechanics. There's a striking and almost suspicious similarity in the way both mysteries protected themselves, and remained unsolved, making people believe the answer was probably beyond them. This 'cloak of mystery' kept many away from looking at the puzzles in a direct, down-to-earth sort of way.

That's because both mysteries gave the impression, to use a technical term, that a bit of woo-woo was at work. Time looked exactly like it was frozen at one level, with all events pre-decided, as if seen from 'the view of the gods'. But at another everyday, human level, events are undecided, with cause and effect, motion through time - our world. The two levels seemed to co-exist in some unknown way. That looked, as you might say, 'metaphysical'.

Quantum theory seemed to imply mind came first, and the physical universe came second, appearing within it. Naturally enough, before the 1920s, more or less everyone thought it was the other way around - the physical universe came first, and mind came second, appearing within it. So time and quantum mechanics both looked like they might go beyond the edges of our science, and off into philosophy, metaphysics, dog-walking, anthropology, and so on. As a result, many physicists kept well away.

So these two puzzles seem to have been preserved by a very particular kind of camouflage: they used the same strategy. This is reminiscent of that bird that stands straight up, very still, on a branch with its beak pointing upwards, looking exactly like another branch, and with just the right mottled greys to stay well hidden. The thing about the common potoo bird - like both of these mysteries - is quite simply that it looks like something else.

So what's behind that, one might ask. Is it just that we're a superstitious lot, so the two puzzles that remained unsolved for the longest were the two that looked like they had an intractable, unscientific explanation?

Or in all the billions of galaxies we can see, are there many civilisations, some of them grappling with these mysteries - which are central to the universe - and which like any good puzzle, contain red herrings to throw puzzle solvers off the scent, make them more baffling, and delay the arrival of a solution? And the puzzles themselves: how random or contrived do they look? If these solutions are right, it's an interesting question.

They're very general. If you make the natural false assumption about frames, then when you get to special relativity, if you do, time looks 'metaphysical'. And the QM puzzle, if my solution is right, looks *exactly* like mind is affecting things, if you start making measurements on particles. But it isn't.

It might even tell us something about other civilisations. If those puzzles look set, then - being central aspects of the universe - they might be set for many, to put it mildly. But to fall into the trap, you need to be open to something more 'metaphysical' about mind or the world, *having seen it elsewhere*. And then perhaps you mistake that, led by the nose (or the whatever), for what is in fact just two areas of ordinary physics. So perhaps tricks like these tell us something about what others are open to, or aware of, in the universe.

There's a bit about the chances of intelligent life on other planets in Chapter 118, and one thing I can say, which as always, only applies if these ideas are right. It's about this principle that kept the underlying physics hidden: having found it in one of them, it helped me solve the other.

When I saw in the late '90s just how the down-to-earth answer for the time puzzle had evaded capture due to a sort of fake metaphysical side, I guessed the quantum theory puzzle would have a similar, purely physical explanation, and searched for it from then on. It was some years later that I read about an idea that came out of decoherence: 'interactions not measurements'. At that point, because it implied a down-to-earth answer, and fitted well with bits of the theory I already had, I felt I was hot on the trail. So if these solutions are correct, that principle was good enough to use as a clue.

112. What exists?

Some things I've mentioned earlier should make better sense now. I hope at least you'll recognise the ideas that led to this view of things, whether or not you draw the same conclusions. I didn't see a lot of them beforehand - there were all kinds of clues that made me wonder why I hadn't seen them, or had taken years to, even with RT right there in front of me.

But the mind often relates to general impressions of things, and is a creature of habit. Like many of us, I was immersed in the standard picture, and once a way of seeing things has got established, it's hard to see a different one. But when a different one appears that breathes sense into the previous one - as this one did, for me anyway - things come out of the fog. Along with the daily plodding process of trying out calculations and ideas over years, the endless backing out of cul-de-sacs, and casting around for new ways forward, there were rare, occasional bursts of excitement as bits of the picture emerged, or when bits of mathematics worked, supporting what was there.

One philosophical thing that came out was interesting. The Greeks struggled with the question of being and change: how something can exist, but change as it moves through time. I don't see that as a problem, but they did. They found resolution in the work of Democritus, who with his teacher Leucippus, got to a hypothesis on the atom. This eased the problem neatly, because the atoms could group themselves into different objects, and while the objects changed, the atoms didn't.

Since then, as we've struggled to understand time, that question has been pushed to a further regress. Particles like the ones Democritus imagined do indeed exist. But they go through changes, it turns out - they're not quite like his picture. Their energy changes when they move, they age at varying rates, they only last for fixed lifetimes, and so on.

And according to rotation theory, the answer to the conundrum of being and change, perhaps surprisingly, is that everything is in a constant state of flux. Nothing *is* - not as we imagine physical existence. With matter, everything is constantly being replaced by similar things, as in the analogy of water from a tap (p 136). So what we think of as existence is emergent, and turns out to be something else. And nothing in the physical world of matter has a more solid existence at all - not even for the tiniest fraction of a second.

113. Wheeler's predicted conceptual solution

From my point of view at least, John Wheeler showed a wonderful ability to forsee what might turn out to be underneath these puzzles. He said: *"Some day a door will surely open and expose the glittering central mechanism of the world in all its beauty and simplicity."*

I used to read the transcript of an interview that Paul Davies did with John Wheeler in the '80s in the book 'The Ghost in the Atom'. I read the book a lot for years, and scribbled notes all over the front and back pages, until one day on my travels, after sitting on the ground reading it for ages at Guadeloupe airport, I got up and accidentally left it there.

It was a series of radio interviews on quantum theory, with some very good physicists from that era, including John Bell. It was great to get *talk* instead of prose, and amazing to see how wide the range of views was, showing how stumped we are in trying to understand the puzzle. That was where Wheeler said that the best hope for progress in quantum theory was *'Finding a new conceptual basis from which quantum theory can be derived'*. I searched for one from when I first read that, and that was thirty five years ago.

But it was only years later that I read the quote from John Wheeler about the 'utterly simple idea', and other similar quotes, and by then I was working on a very simple idea. To me it seemed wonderful that he'd guessed there was a simple conceptual solution.

The theory wasn't in shape to show anyone until around 2011, when most of the mathematics came out, but I didn't know that at the time. Without really knowing how I would approach him, I very much wanted to go and see him with what I've been describing to you, and ask his opinion.

114. Interdependent concepts

I mentioned that there were three lateral jumps that led to the theory. They should make sense now. They're the background rotation, matter as waves in the dimensions, and the replacement principle.

Taking matter to be waves in space isn't revolutionary: what makes it lateral is taking matter to be part of the small-scale dimensions.

These three ideas are very interdependent (there's a bit on interdependent lateral jumps in Book I). Without the background rotation, matter as part of the dimensions doesn't necessarily help. But without taking matter that way, the background rotation would 'spin its wheels' pointlessly. So the first two ideas need each other.

But without the replacement principle to cause matter's time rate variations, either of them still looks superfluous, and even both together doesn't make a lot of sense. But taken on its own, the replacement principle does f(-) all.

But there are really four lateral jumps, or perhaps three and a hop. Without the Newtonian level of time, the replacement principle is useless, as it needs time to work. Even if one does reach it, it seems irrelevant: without another level, which - if you understand physics well and still put it in - is also a lateral jump, the other three ideas solve nothing.

But without those three, it seems pointless and extravagant in assumptions to put in the Newtonian flow. Only when all four ideas have been arrived at (which can happen if you just bash away at it for years), and are put into the picture in the tightly interlocking way in which they can be put into it, does a viable solution appear.

115. What false assumptions need to be dropped?

Assumptions sometimes need to be thrown out. Breakthroughs often involve letting go of assumptions. There are also general ones that keep recurring. A key one to boot out is the idea that the present picture is almost complete. That gets passed to students by implication. And it's not just nowadays - it happened in the late 19th century as well, as Max Planck found out. Although it didn't stop Planck, the idea that what we have is nearly complete is hugely damaging - it shuts down the use of the imagination in a major way, and has been an absolute disaster for physics in the 20th and 21st centuries.

Another false assumption to be dropped is about time, it's the idea that 'we can't have two levels of time'. The answer to that is: we already have. And in this alternative view it's much better, as it's not two levels, but one level and a mechanism.

But I've found another very general kind of error, and it turns up all over the place - in both physics and philosophy. And that includes in my own thinking: looking back with hindsight, I find it again and again.

It's that we assume *connections between things* that are in fact unconnected, or connected up differently. An idea is taken to come with baggage, or other ideas attached to it, when it might not include the connections we assume. People say to themselves: 'oh, but if I believed that, I'd have to believe this, that, and this as well'.

It's not true. I've learned that an idea needs to be taken initially on its own. It should not come with pre-assumed connections. For now, the pizza is just a single pizza, normal price. It's not part of a two-for-one special offer, or some other price reduction system that connects it to other pizzas. The reason we shouldn't assume links before we start, between one idea and ideas around it, is simply that the other ideas might be false, or unconnected.

And if we take ideas separately, at first anyway, we can look at the idea of a transmitting medium for light, without assuming it's the one that was ruled out in 1887. We can look at the idea that how matter is *viewed* is important, without assuming the observer has anything to do with it. We can look at the fact that Newton's gravity theory works very well, without assuming it does so in all areas. The same with Einstein's. And we can look at the point that experiment has shown special relativity to be correct, without assuming that therefore Minkowski spacetime is as well.

And that brings me to pairs, or combinations, of theories. Many people think both parts of relativity are right, a far smaller group think both are wrong. An even rarer view - during the 20th century - was one right, one wrong. It was also rare to think special relativity right but spacetime wrong - the common views were both right, or both wrong.

But these patterns have shifted all over the place in the 21st century. General relativity has been demoted to an approximation, spacetime to an emergent phenomenon. But among the many who think general relativity is ultimately wrong, only some think curved space is wrong - many believe the final theory will include it. The view from RT is: special relativity is right (in almost every way), spacetime is wrong, general relativity is wrong.

And the idea that the early part of relativity is right, but the later part wrong, would explain a lot. Special relativity has combined well with other physics. Fusing it with quantum mechanics led to QFT and QED, wide ranging theories that were beautifully confirmed and very successful. But all attempts to fuse general relativity with quantum theory have had no success, and have been stuck for decades. General relativity has not fitted well with anything else, and fitting well with anything else is a good indicator.

So initially anyway, we need to take ideas separately. If you look at the ideas people had in physics a few centuries ago, they were a mixture of true and false, just as ours will turn out to be, when our great grandchildren look back at the ideas we have now.

116. Outsider physicists

Talking about psychological aspects of physics, it's worth mentioning one or two other things - we've learned a lot over the 20th century. The points that follow are just loose patterns that sometimes appear, as one strand among many. Nowadays we know there can be *an almost innate resistance to truly new ideas* in science, even if they might be right.

There are a lot of examples of outsiders with apparently 'crackpot ideas' who turned out to be right. Phil Gibbs has compiled a list of 'crackpots who were right' - there are excellent physicists there, including Nobel prize winners like Chandrasekhar, and many who couldn't get through peer review or struggled to publish, and fought against a tide of resistance. Nowadays good physicists such as Lee Smolin are sometimes called crackpots on the internet, but some use the word simply to mean *'a physicist I disagree with'*. Lee Smolin is well respected, and he has had the honesty and courage to question the standard view in physics. The thing to remember is that in the past, every established standard view - without exception - has turned out to be wrong.

And as I said, even Einstein was an outsider, being an amateur physicist with no academic affiliation when he published special relativity. Physics has been changing in the 21st century, we have open peer review and other systems, and there's a new recognition that physicists should be judged by their work, rather than their background, or anything else. It's also understood that the ideas of some outsiders will be valuable, as they get a different viewpoint on the landscape, and might see what others miss. And less formal training can mean less immersion in the previous set of ideas.

But although things are changing fast, and we don't like discrimination these days, it's amazingly bad. Many journals don't allow anyone to submit a paper without an academic affiliation. Independent researchers have to publish in lowly journals, and are then ignored or criticised for doing so. You've got to have a teaching job at an institution. So Einstein's historic amateur paper *'On the Electrodynamics of Moving Bodies'* would be thrown out, and not even read. If Einstein tried to submit that paper to such a journal nowadays, he'd be unable to fill in the form. And yet during his 'annus mirabilis' - his year of great achievements, amateur papers flew, and changed the world.

But the reality is, the outsiders who have made vital contributions to science are just the ones we've heard about. They're a very small minority who were lucky enough to get through. And of the 'famous three', Galileo, Newton and Einstein, TWO were outsiders, of one kind or another. Galileo 'dropped out of college', changed his subject to mathematical physics, then taught himself. (He eventually worked at Pisa University.) But those kind of scientists are just the tip of the iceberg. Many others didn't manage to publish their ideas, or had them stolen by a mainstream that is in a better position to publish, and will sometimes harvest the creativity of free thinkers, locking them out, and taking the rewards for themselves.

There's that well known story of the high-roofed lorry that got stuck in the entrance of a tunnel in the alps. The firemen couldn't tow it out, people got out of their cars to watch, and a five-year-old Swiss girl suggested the lateral solution of letting its tyres down. What makes the story unique is not that an outsider saw what the experts didn't - it happens a lot. What makes it unique is that we've heard about it at all. The common pattern of events would be that someone else takes the credit.

And in physics, before new ideas get taken seriously, they're often resisted, for a number of reasons. They can sweep away or devalue existing work, or just the time spent studying existing work. And some get a sense of security from the 'solidity' of the standard view.

There's a quote attributed to various people, including Mandelbrot: when a radical new fact appears, reactions tend to follow the sequence: *1. disbelief, 2. it's contrary to orthodoxy, 3. we knew it all along.* The good news is that nowadays we know about this pattern. We also know about the closed shop, and the discrimination. We're onto it, and change is happening.

A resisted idea, of course, isn't necessarily right - bad ideas are resisted too. But if physicists get into two groups about a new idea, and argue about it - some for, others against - it could be right. One group might be against it for the wrong reasons, the other could be being more honest.

Among the rebels in physics, some challenge the standard view for irrational reasons: some have an antipathy to authority. But there's also another kind of rebel - the kind Einstein was, who genuinely wants to know what's out there, and questions the existing view because he can see beyond it. I don't really fit any category - I rarely see myself as a rebel, more as someone who quietly points out overlooked clues, coming from the overlooked conceptual side, who where he disagrees with others, ignores it wherever possible. And when I talk with full-on rebels, if they don't know of my ideas, they take me to be rather mainstream - they sometimes even rebel against me! I *always* find myself defending aspects of the standard view, such as special relativity (amateur theory though it is), the experiments that support it, or the general scientific method of going by experiments.

Anyway, with psychology in physics, points like these don't necessarily help, and human behaviour is often far, far too complicated to generalise about. And the only way to test theories is, as always, by experiment, rather than by expensive psychoanalysis of those for and against. But it's worth being aware of points of this kind, nonetheless.

117. Back to that experiment

Talking of how experiments are the only way to tell what's going on, I can go back to the 2015 experiment done by Batalhão, Paternostro and friends now, outlined in Chapter 58. They found an unexplained direction of time at the quantum level, and an irreversible world down there - right where the world is thought to be reversible. But that reversibility, now shown to be false, was what led to wide areas of the standard view of time.

But another view of time explains it. In rotation theory, time and its direction start at a *smaller* scale than the quantum level, not at a larger one. Down at the Planck scale, a direction of time is created by the replacement process. So it causes the direction of time at larger scales, including the unexplained one that was found at the quantum scale.

Is the replacement just another one way process, providing an arrow of time, at a small enough scale to explain the experimental result? It's more than just another irreversible process. It does make a new arrow of time, though without block time, the desperate need for one we're used to, much needed to patch things up, evaporates.

But because it relates to how matter is created, has its existence, and moves through time via the replacement principle, this particular arrow of time runs very deep. It's not at the deepest level of all, but what we mean by the word 'time', going by Einstein's definition, is *what clocks measure*. In those terms, the background rotation, and how it causes the replacement process, is the very source of time itself.

118. Time travel

People wonder a lot about time travel: it's a fascinating idea. The question is whether it's possible in principle, for an advanced civilization. I'll give you my thoughts on it, some of them more speculative than others. According to this theory, you can go forward in time but never back. We all go forward in time anyway, but to travel centuries into the future, it's the same as in standard theory. You use time dilation, in one of its two forms.

It's either about travelling away from the Earth very fast, then coming back, or hanging around a strong gravity field for a while. Those two effects both create a slow time rate, so on your return, Earth would have moved on. But to get a significant shift - to reach high speeds, or orbit near a strong gravity field safely - you'd need advanced technology.

High speeds is a better bet (I'm just saying, that's all). At Sgr A*, our galaxy's central black hole, if you orbit where standard theory predicts the innermost stable orbit will be found, where the inner edge of the accretion disc should be - although observations show this isn't true - your time rate is only slowed by about 0.8. The orbit speed slows it by 0.9, those combined take it to 0.72. So for every year there, 17 months passes on Earth: not enough.

If PSG is right, you can improve on that a little, and get a bit nearer the black hole than this limit from standard theory - but let's not go there. Far better, if we could travel at 0.99*c*, 7 years passes on Earth for every year onboard the ship. At 0.999*c* it's 22 years, at 0.9999*c* 70 years, and it goes on, and *way* up. So if speeds like that are possible, so is serious time travel.

One can get speculative. I should first let you know that in my view UFOs are not spacecraft. Nevertheless, there may well be other civilizations, in all the billions of habitable planets we've just found out are likely to exist, by finding a lot of planets right nearby. That's only since '95: before that we didn't even know if there were other planetary systems like ours. So now there's a lot of room for life, and intelligent life. And recently we found out that intelligence has developed independently at least three times on this planet: dolphins, elephants, us. Forty years ago that wasn't clear, but even crows are looking smart nowadays - and they're sitting on a different branch of our family tree. So intelligence happens easily in the universe.

Incidentally, dolphins and elephants have a sense of humour. So among all the amazing things that might exist out there, humour is likely to be around the universe, and a lot of it, which is cheerful news.

This possibility of going forward in time, but not back, is a basic aspect of the universe if RT is right. One can try to guess what it might do. For one thing, it might mean advanced civilisations, if there are any, would need to be very stable for their travellers to choose to explore the galaxy at the necessary speeds, because they'd keep returning centuries on. If their civilisation, being older than ours, had got rather stable, so on their return the place is going to be much the same, that would help.

What about the past? It's possible, in principle, to create a 'time viewer', and watch the world in 1850. The light from that era is still travelling outwards, and if it reflects off anything out in space, or swings around a black hole and is sent off at many angles, it's not impossible, *in principle*, to get a look at it. But you can't actually interact with people from that age.

The nearest thing would be to reconstruct them, which would be no good, as what they went through shaped them: nurture, as well as nature, is a large part of it. But it's the best we could do. The little circles of waves they were made of no longer exist - not in those patterns, and nor does whatever else made them what they were.

This doesn't affect ideas such as reincarnation, but whatever one believes, those particular people have gone. And so has the period in history, except in what it leaves behind in other ways. And what it leaves behind in other ways, though not enough for Woody Allen, would just have to do. He said: *"I don't want to achieve immortality through my work; I want to achieve immortality through not dying. I don't want to live on in the hearts of my countrymen; I want to live on in my apartment."*

119. The unreasonable effectiveness of mathematics

I can describe the overall picture that comes with the theory now, having put as many pieces of the jigsaw in as I can for the time being. In this picture, the world we know arises from side effects of a simpler and more fundamental process. This deeper process happens at a scale so far away that there are no obvious clues to suggest it's happening. Matter seems solid, and to have an existence in its own right. And yet all the patterns we know in the large-scale world are in fact written onto the background rotation, so they're secondary patterns - that is, a pattern written onto another pattern.

Standing further back, one sees an oddly simple, unfamiliar picture. There is only the background rotation. Round it goes, and all the structures of matter arise from patterns made out of slight variations to its smoothness. And the picture of gravity depends on another set of patterns, all made out of slight variations to its speed.

All these patterns are from tiny standing ripples and irregularities in a fast-moving stream, which seen from some other point of view might be utterly insignificant. But they're varied enough to make our world. The regularity of the stream itself, on which everything is a secondary pattern, explains why the world of matter is so full of regularity and symmetry, and is so consistent and predictable.

Particle physics is full of subtle and less than subtle symmetries. One particle resembles another, families of particles have family resemblances. And the order and symmetry that is found in many areas suggests a common cause at the root of things, with some kind of deep regularity, which is somehow in a position to shape everything else.

And even more generally, this makes it easier to explain what Wigner called *'The unreasonable effectiveness of mathematics in the natural sciences'*. In a paper with that title which became very well known, Eugene Wigner argued that mathematics is unnecessarily good at describing the world. Wigner was given a Nobel prize for work on particle physics, including the discovery and application of fundamental symmetry principles. He saw that matter makes patterns. The universe is so predisposed to mathematical description, that it has often been taken to need an explanation.

Attempts have been made to explain it in various ways, and in 2015 the FQXi essay contest, an event in physics each year, was on the question. The range of suggested answers is wide every year, because the set question is always a controversial one. But that year it was even wider than usual.

As with many problems, this one about mathematics is best looked at with the simplest example you can find, to get to the basics. I used to think about a see-saw. Put two objects of different weight on, at different distances from the centre. That's four different numbers. And you find there's a simple rule: when they're balanced, weight x distance = weight x distance. Why?

It's a universal rule: it holds for a see-saw made of any kind of matter, and is simple and basic enough to make one wonder what's behind it. Why should multiplying those four numbers in pairs do that?

Intuition tells us that distance from the centre makes an object heavier, from a balancing point of view, and increases its effective weight in proportion to the distance. But weight and distance are very different things, and it's not clear why they should be linked in this way. Weight is only a step away from inertia, which we don't understand at all well. But it seems that the world is very 'linked up', for whatever reason. It's easy to take that for granted, but Wigner didn't, and nowadays thinkers don't.

Rotation theory has a new kind of explanation. The surprising usefulness of mathematics can be seen as the result of the background rotation, rolling on underneath everything. It gives rise to all the different kinds of matter, so its regular speed, and the regularity in various other basic parameters, lead to a *standardisation* in matter. (It's like a bubble blowing machine: one could be built that blows bubbles of a very standard size, because of its nature.) That's what causes all the order, simplicity, self-similarity, and hence amenability to mathematical description, that we find in the patterns that matter creates at larger scales.

Could John Wheeler have unconsciously seen a glimpse of the background rotation, spinning regularly underneath everything, when he decided to use the word 'glittering', to describe (possibly referring to a vague image of little spinning wheels) what he talked about as the 'glittering central mechanism'? He thought about it a lot - a very simple, central idea behind all of physics. So perhaps it was somewhere in his mind's eye - it's possible.

Part 26. Conclusion, speculation, and a few stories

120. Both sides now

If rotation theory holds up, and goes on providing a good set of explanations, it will show the importance of the conceptual side of physics. So, as I've said, if someone sets out a more complete version of the mathematics, they can't take the credit for the theory. In the past, attempts have been made to take the credit for other discoveries by doing further work on the mathematics, as if any discovery is only the mathematics of the discovery.

But special relativity was a new interpretation for mathematics that basically existed already - the Lorentz transformations. Still, Einstein should certainly get the credit. That applies although other mathematics was done on it, both before and after his contribution. That's just what always happens.

RT is partly a conceptual picture, and if it's right, it would shift the emphasis in that direction. Interpretations are vital, and lead to new mathematics if you follow through. They were taken by a few to belong outside the field, but only because we were having trouble with them. And by the 21st century there were calls for a return to the conceptual side.

Standing back, the path we've been on for centuries, is of unifying physics. What will physics unify *into?* Many think it'll be more than mathematics. The answer will show what the question was all along. If Wheeler and Einstein were right, the last step is conceptual. If so, the goal will have been a picture *and* a mathematical description all along.

I've repeated this point because it means that these areas I've covered may be needed, even if RT seems to lack mathematics in places. In fact, often the mathematics had been done already, but needed an interpretation. Work on the theory is ongoing: so far special relativity has been reproduced, in both the phenomenology and the mathematics. But with quantum theory, only the phenomenology has come out so far. Others may get to the mathematics before I do.

The set of ideas that anyone would have to use to do that, DQM, took a long time to assemble because of having to rule out similar, but false, conceptual pictures. To pick out the one that worked best in each area took years. Some lateral solutions look obvious once they're seen, almost as if we knew about them all along. This might hide the fact that the dependable landscape I've set out here had to be fought for, often every inch of the way. I hope it won't be taken for granted.

But the gravity theory produced new mathematics. I delayed for ten years to get there. Where existing mathematics is right, you just try to get to a picture that goes underneath it. But if you think it's wrong, you try to replace it. And if you get to mathematics that is similar, but a bit different, then the picture can be tested.

And if you find the right picture, it might be possible to prove it. This has not been done before, but if this one's right, no-one has found the right picture before. RT includes a proof, or strictly, a near-proof. It shows in a simple way, beyond reasonable doubt, that every point on an orbiting object's trajectory is connected to every other point, via the law of refraction. Or it can just be a falling object. And that (even if some physicists go rather quiet, as they can't pull it apart), proves that refraction causes gravity. This supports the entire theory - it's in Appendix D - because physics from different areas of it went into that bit of mathematics.

121. Unfinished ideas

Well, that's about it, and I hope it has made good sense. I'm not going to go far into speculative ideas, but in two areas the ideas have solidified recently: the vacuum catastrophe and the early universe. The questions there are so important, they justify telling you what I've got on them. And I must tell you about MOND, as I've recently found a way to explain it.

But there are always new ideas coming out of the previous lot, and there has to be a stopping point, where things get wrapped up. And going by what I've learned on the way, until something is truly explained, it isn't. Loose ideas that look promising often lead nowhere. An idea for which I haven't got any mathematical confirmation, or cross-corroboration with more dependable parts of the picture, starts to look suspicious quite quickly.

But physics is a joint effort, even if an individual can get stuck in a timewarp (when I came out of years of work on the theory, my friends had all moved on, and their children were doing the same). Over long periods, it's always a group activity - any breakthrough is made with the help of past and present contributors. Between us we steadily carve a path through the unknown, and towards a better understanding. The group aspect is the bit that matters, so you have to stand back and look at the whole thing.

The way the standard model was put together shows it - it was a group effort by many, one of the real achievements of the 20th century. By then we were working together as a species. The International Space Station shows it too, and best of all is the spirit of cooperation and togetherness between people from different places it has inspired. That spirit is appearing in many areas anyway. It's not to be underestimated - there's a lot of division in the world, then there are things that bring us together. Science is one of a list of them, that break down barriers, and make us feel part of something good. As well as the troubles this century, we also have that new global spirit of everyone connected up and together, no matter who you are.

Because whatever I've done on this jigsaw is part of a group effort, clues only half understood may let others take it further. So where there's what looks like a good clue, or a few jigsaw pieces that really do seem to fit together, I'll share them, and leave a signpost pointing down an unexplored path, like in that book. What follows includes unchecked ideas that I'm not certain about. They're definitely not to be taken as central parts of RT, but some areas are spinoff theories, and can just be taken as they are.

122. The other dimensions

To add something to the last chapter, recently a possible goal for the future has appeared. A planet in the habitable zone, near the size of Earth, has been found 4.2 light years away, at Proxima Centuri. It's amazing to find one so nearby that can support life, and interesting that it may be a goal beyond the solar system for future adventures.

There's so much that needs taking care of on Earth now that any goal beyond it looks irrelevant - we need to look to home before looking anywhere else. But perhaps one day sending robots or people there will inspire unity as the ISS has, with the shared adventure of exploring further. Some day, Proxima Centuri b may be what the Moon was to the 20th century - a reachable first step, that gets us on our way.

The theory I've described works well using four basic space dimensions, with Newtonian time, or something like it, as a backdrop. With that as a starting point, it reproduces the phenomena and mathematics of special relativity, also the phenomena of quantum mechanics. And it closely mimics general relativity, producing different mathematics. It does those things via just one rotating circular dimension and three flat ones.

Some might say general relativity also describes the universe well with four dimensions, including time, and that it's more economical. But RT, using four dimensions *plus* time, can explain a long list of phenomena in physics. Some explanations are more complete than others, but they all cross-corroborate each other, and fit existing mathematics well, and many of the things they explain have no other explanation at all.

So using four space dimensions, and deciding time isn't a dimension, one can explain a lot. But there's still the possibility that there may be more circular dimensions, and that the others are rotating as well.

If so, the reason the universe can be described in this simple way, with four dimensions and time, is that in the geometry, only one rotating dimension is needed *at a time*. That is, in very many situations. But elsewhere in physics there has been some suggestion that different circular dimensions may carry different forces, and this is potentially compatible with rotation theory. And what I've said about Kaluza-Klein theory is exciting, and may show the future of physics. But although it's possible that the final version of RT will involve a number of rotating dimensions, either way, for now the universe can be well described using just four space dimensions and time.

123. The vacuum catastrophe

There's a problem that some see as the largest problem in physics. Whether or not it's the largest one, it's a major one. It's a discrepancy between theory and measurement, and between two separate theories. Neil Turok once told me he saw it as the main problem facing the next generation of physicists. I told him I had a solution, but it was a filmed conversation with limited time, and I hadn't told him the background theory.

And you need the background theory to explain it. That's why my solution is here, and not in the gravity book, Book II. But although people usually don't say this, the vacuum catastrophe is partly about gravity. One element of it is a catastrophic failure of a gravity theory called general relativity.

The discrepancy is a puzzle for which there's no explanation at present. Two ways of estimating a particular number come out different by a huge factor: 10^{120} (a 1 followed by 120 zeros). What it shows is that somewhere, there's something seriously wrong with the standard picture. And the problem turns out, yet again, to be a conflict between relativity and quantum theory. One of them is likely to be wrong here, and I find quantum theory is always right. General relativity is always right too - in the areas people talk about.

But that's changing, and taboos have been broken recently. As a reminder of the quotes I've made already - and they're important, as no-one dared say it a few years ago - Raphael Bousso has called general relativity a large-scale approximation. That means wrong. Sabine Hossenfelder has said it's a large-scale approximation, and that most physicists think the same. Many assume they know what will replace GR, but if you have nothing to do that, you can't assume *anything* about what will replace it. We've been trying for quantum gravity since the 1930s, with no success.

Nowadays some openly criticise GR, others say it has 'never been wrong'. But the vacuum catastrophe is one of the places where GR inevitably does look wrong. It's an area in which GR and QM disagree, and it's clear enough that quantum theory wins.

Quantum field theory, QFT, is one of our centrepiece theories. It comes out of QM, and works exceptionally well in almost every way. The only doubts on it are about where it doesn't connect with GR, but then a lot of things don't. The predictions of QFT have been confirmed again and again.

The question is, how much energy does the vacuum contain? How much per cubic meter? That's the number we can estimate in two different ways. For the 'bubbling soup' of the vacuum (Chapter 76), QFT predicts a large vacuum energy, from all that bubbling that goes on. And reassuringly, the numbers it gives are supported by measurements made in the '90s of the Casimir effect. Loosely speaking, we've measured the bubbling. So a well-established theory makes a numerical prediction, every other prediction it has made has been neatly confirmed, then experiment confirms this one as well.

So far so good, science is working, quantum field theory is right on track and self-consistent. But when we look at what GR predicts on this, then compare it with another way of estimating the same number, we get a discrepancy so enormous that it's hard to know whether to laugh or cry.

The other way to measure this is to look for the *gravitational* effects of all this energy we know is there in space, out in the wider universe. We do this using our theory of gravity, GR. Energy is equivalent to mass, and GR says all energy creates gravity. It's an important part of the theory, and if it fails, GR fails. In this instance, GR leads to the idea that there will be a kind of 'reverse gravity' - an outward pushing effect. An outward pushing effect would mean there's a cosmological constant.

From the late 1920s to the 1990s, people thought the cosmological constant probably didn't exist. It had been a fix-up factor, a mistake of Einstein's, that we might never need again. Then in the '90s some measurements, including the supernova data that led to the idea of dark energy, suggested a positive cosmological constant, but a small one. It was far too small to be the same one as the (reverse) gravitational effect of the bubbling soup. Looking at the way the universe is expanding, which we know at least approximately, if you try to calculate the maximum outward push that could be adding to it, it's not enough. It's 120 powers of ten too small.

This strongly suggests that general relativity is wrong. Quantum theory never misses, and some say GR never misses either. But if we're honest, it misses here. After all, when we look for the vacuum energy *without* using gravity as in GR, we find it, numerically confirming it by experiment. But when we look for it using gravity as in GR, we can't find it.

Even in the solar system we can't find it. If this extra gravity existed it would have a huge effect on the orbits of the outer planets. So a definite prediction from GR has turned out to be wrong. The only way out for GR is if the energy doesn't exist, so the gravity doesn't either. But experiment, matching up with a theory that really has never been wrong, QM, both say it does exist.

Now negative gravity has never been observed, so perhaps we shouldn't be utterly surprised if it doesn't appear. But that's still a failure of GR. And GR, or someone, still needs to explain why we find *neither* negative nor positive gravity coming from all this energy. We've detected it, there's a vast amount of it per cubic meter, and being energy, GR says it creates a lot of gravity. So it's not just that someone needs to explain why it doesn't create that gravity: any gravity theory worth its salt should explain it.

Looking at PSG I'll start with why there's no negative gravity. In PSG, negative gravity doesn't exist. To get an outward pushing force, a mass would have to emit a refractive medium that's graded, but in the opposite direction. That is, getting increasingly dense further away. But the 'gradedness' of the emitted vibrations arises because they dissipate as they move away. That goes one way only, so there's no negative gravity in PSG.

But there's a more general question: why doesn't all that energy create any kind of gravity? It's lucky it doesn't, as it would swamp all other gravitational effects, and destroy the balances that keep us alive. So it's not that we *mind* there being no gravity from all this energy - (just to avoid misunderstanding, we're not complaining in any way at all, or asking for it to be switched on) - it's just that we want to know why.

In RT, the bubbling soup is brief disturbances in the background rotation. It involves activity, so like all activity, it creates resistance to the flow. So it has energy. In RT, the energy exists, which fits the fact that it was measured. But does energy always create gravity? It does in GR.

Gravity in PSG, instead of being caused by curvature, is caused by a gradient in the transmission speed of space near a mass. Now if the zero-point energy is the same right across the universe, as it's thought to be, and if it arises at or near the Planck scale, then all those bubbling soup particles have an effect that might be evenly spread. If it happens evenly enough, it doesn't create a gradient - and no gradient means no gravity.

Instead, all the bubbling of the soup does is it alters the overall transmission speed of space, slowing it down slightly everywhere. What it actually slows is the average transmission speed. But c, if you look at the collisions at a very small scale, is probably going to be an average speed anyway. So without the bubbling soup, the transmission speed would be slightly faster.

The vacuum energy affects space evenly, without a gradient, as part of the background. So it affects the flow everywhere, not just in places dotted here and there. This leaves the universe much as it was.

Physics with a different background speed is almost the same, though some numbers, such as the masses of particles, are very, very slightly different. All that happens is, the gravity of the clumped masses in the universe appears on top of this slightly different background speed. And that means that any difference to the force of gravity, due to a slightly slower background speed, is *already contained in all our numbers* - for instance, in solar system orbits. So that's why we don't find it: it's already in.

So what we call the speed of light would be the transmission speed of empty space, same as it ever was, but after this slight slowing has been factored in. This is a good, viable solution. And there are not many other viable solutions out there, if any. The solution from RT is that a 'background' of evenly spread vacuum energy disappears - into the numbers we already have.

But there's a question about what 'evenly spread' means, and at what scale it starts. Is a gas evenly spread, with its drifting molecules here and there, or dust particles? Very unlikely: gas and dust particles have their own miniature gravity fields. They make a gradient within the transmission speed, so they're part of the foreground, not the background. It's probably at the Planck scale that this appears. If the effect is evenly spread at the scale of the background machinery itself, it gets built right into it.

So this looks like one of the differences between GR and PSG. One gravity theory predicts what we find, the other doesn't. The question arises, can an experiment be devised to test for the root cause of this? Most Planck scale effects are out of reach - from direct experiments. According to RT there are plenty of indirect effects at larger scales, and rare exceptions like the Casimir effect, which provide the occasional direct one.

So you never know. With later 21st century technology, perhaps we can find differences in the way the Casimir effect behaves. Or perhaps a new indirect method will turn up - anyway, it's early days.

124. The early universe: what the JW Telescope might see

The James Webb telescope made it to its place of orbit this year. It's in many ways the most advanced experiment ever created, and it means we'll be able to look further than we ever have. It's there to help us push the boundaries, and try to find out more about where we are, and what we are.

The two main reasons it was given so much priority, and worked on at such a high cost for so long - 25 years - are quite well known. But the urgency of one of them is less well understood. The first is to look for life by studying planets beyond our solar system, which we've recently found out are everywhere, around many nearby stars. It's exciting, because the spectra of the light that comes through the atmospheres of these planets can be analysed, and they may well contain signatures of life.

The second is to look back to the early universe, to understand something very important that has baffled us for decades. When we look at the furthest objects, we're looking back in time, at the early universe. We expect to see what our theories tell us we'll see - a forming universe. But instead, what we see is an already formed universe.

The 'Hubble deep field' is one of the most famous photos ever taken. But how it was taken is a great story. The James Webb's predecessor, the Hubble Space Telescope, was run from the ground via the Space Telescope Science Institute, directed by a man named Robert Williams. He had what's known as 'director's discretionary time' on the telescope, and at one point he decided they should use all 100 hours to gather light slowly, pointing the best tool we had at an apparently empty little patch of sky.

He was advised by colleagues not to do this. They told him the image would be black, he would see nothing, and waste telescope time. Theories of galaxy formation predicted that beyond where we had looked already, there was nothing to see, as nothing had formed yet. But he wanted to look anyway: to look as far as we could look, and see whatever we could see.

The result was one of the most important images ever taken - out of one tiny corner of the sky swam three thousand galaxies, of all shapes and colours, further and deeper than we'd ever looked before. Each little patch of sky the same size was likely to have about as many galaxies in it.

That was the mid '90s, and since then we've looked further and further back. And we keep on finding the same thing: not a forming universe, an already formed one. I'm not saying the big bang never happened - I don't take that view. I'm just saying that the enormous push of time and money that led to the James Webb telescope was driven by a need. It's up there to settle a very urgent question, because we need to look even further back, find out what's there - and if we can, to watch the universe getting started.

125. An error about cosmological time dilation

What's known as cosmological time dilation is not really time dilation. It's an observational effect, and not like time dilation due to motion or gravity. The two real forms of time dilation, due to motion and gravity, can make objects get out of sync with each other, and have age differences when brought back together. That's why they're real: they have real physical effects.

Cosmological time dilation, by contrast, is a kind of illusion. It arises because waves get stretched on their way to Earth over very long distances, making observed events seem to happen more slowly. It's about what we see, not what exists, so it can never make objects get out of sync, and there are never real physical effects. For an analogy, you can't touch a rainbow: it's simply an observational effect.

Some science articles fail to mention the difference, which leaves the public confused or unaware - this happened when cosmological time dilation was observed in quasars a while ago. There are reasons: some writers don't know the difference, or may leave it out because it makes a better story if the time changes seem to be real. Some don't mention that there are real and unreal versions, as we'd then have to say more, and it's embarrassing how little we know. And some may have read something misleading about it, as seems to have been the case with the observations of quasars.

When this happened, bad sources got it wrong, better ones got it right. One respected physicist, Sabine Hossenfelder, said *'Cosmological time dilation is purely an observational effect. It's about how events from back then look to us'*. And Ethan Siegel, a physicist and science writer, said: *'In a sensationalistic and quite incorrect fashion, some outlets have been reporting that this means "time ran slower in the early Universe," which isn't right.'* He also said: *'That's not how any of this works'*.

It may help to show how sources can mislead. Further down the same page, they may make it clearer, and say that time doesn't *actually* run slower. But there's so much confusion over time dilation anyway that this may not help, particularly if the headline has misled the reader earlier on. In one, the word 'seems' was used in an ambiguous way. An article headed 'Cosmological time dilation', had the sub-heading: *'Time seems to have ticked more slowly when the universe was young, according to observations of ancient astronomical objects'*.

Now this, as you can see, has two meanings. If you saw it, you'd think it's one thing, but it's actually the other. It looks like it must be saying: *'It seems that time ticked more slowly, from observations...'*, as if new observations suggest time really did. But the real meaning is *'due to an illusion we understand, one gets the impression from observations that time ticked more slowly...'*. So it may be accidental, but the word 'seems' can be misleading. I saw it in the print version - the online version was I think edited later, and improved a lot, changing it to the word 'appears'.

But as you can see, scientific information can get distorted on the way, just as waves travelling across the universe can get stretched out of shape.

126. What other clues are there?

When we look back as far as we can, the main thing we see is very rapid star formation. Galaxies formed unexplainably quickly, immediately after the big bang. The strange thing is, the large ones formed quicker and earlier than the small ones. This pattern, known as 'downsizing', goes across billions of years: galaxies that form get steadily smaller, and form slower. It's the opposite of the hierarchical build up that our theories tell us we'll find.

It almost looks as if time was running faster back then. Now we know beyond doubt that in *some* situations time runs faster than in others. It happens in a real, physical way. We know two effects, time is a mystery - three is possible. But we have *nothing* that could do this: no existing view of time has a reason for an overall time rate to run faster earlier on. But if it did, it might neatly explain how galaxies had time to form so early, and to already be there.

And some of our present ideas look strikingly like an incomplete picture of a faster time rate. There's a well-known theory that light travelled *much* faster in the early universe. VSL (varying speed of light) theories have gone from a heretical outsider's view to an acceptable one, in about ten years. One of the problems it can solve is the 'horizon problem' - how did different parts of the universe get so similar, in temperature and in other ways? Lightspeed is too slow to get the information around.

And then there's inflation. A very rapid burst of expansion, far faster than the physics we know, in the universe's first moment, is found to explain a range of problems and issues, and seems to show the origin of the structure of the universe. But we have no mechanism at all, and the theory has many critics, and huge problems. Paul Steinhardt won a Dirac prize for his contribution to inflation, but later because one of its main critics, seeing it as so flexible, that it can be made to fit almost any data.

The Planck satellite results in 2013 brought problems. One, the 'unlikeliness problem', is that after a period of inflation, the odds that a universe like ours will arise are one in a huge number: a 1 followed by 10^{100} zeros. That's not an explanation! The data shows us that in the real world, inflation doesn't work. Penrose says inflation has been disproved anyway.

But VSL and inflation showed - loosely it must be said - that something like a faster time rate in the early universe could explain a lot. So how would that work? In rotation theory, the time rate is set by the speed of the background rotation, because it's also the speed at which matter is being replaced. Is it possible that the background rotation started very fast, then steadily slowed down? I've wondered about that for fifteen years, and recently came back to it, as a new idea has made it look a lot more possible.

But what I have on it so far is comparatively speculative, and it's not part of the main theory at present. It's a spinoff theory that has started to look good enough to write down. But that means pretty good.

127. A possible cosmology

The new idea I had was during an hour with nothing to do. It was too hot to walk far, I was sitting on a low wall in the shade near a parking lot, behind a building in Arizona. My partner was at an appointment, I just had to wait.

To me it was a breakthrough, although it might not seem all that interesting - but that's why I mention where I was at the time. Until then, ideas about a varying time rate cosmology had been hard to get to grips with, and nothing certain or reliable had appeared. But over many years, the idea occasionally solidified. The stages it went through won't take long to tell, as it hardly went though any. It may have started as early as 2004, and was one of many ideas, most of which led nowhere, and got shelved or thrown out. This one didn't even have a mechanism for the first ten years.

I knew friction wasn't a viable mechanism for the slowing of the background rotation. I've found no evidence for friction at the Planck scale. A planet's gravity field slows the background rotation near it, but it speeds up again at that place in space if the planet moves away. This is like light speeding back up after being slowed, which it does. The background rotation behaves like light. It goes by the rules for wave motion, travelling at the local transmission speed for waves, as light does.

The flow slows if the local transmission speed changes. So the best way for it to slow everywhere is if the *general* transmission speed changes. That would make sense: the background flow travels at the present transmission speed, c. In an earlier era, could c have been faster, as VSL theories suggest? If the background rotation was faster, so time was as well, lightspeed would also be faster, compared to this era.

Materials is not my field, but one day in maybe 2015 or so, I read about how many materials conduct waves quicker if they're more rigid, slower if they're less so. There are plenty of reasons to see an analogy between space and a material on Earth - I've mentioned them already. At that point, some years before coming back to this, I realised that as space expands, it might behave like a material that gets stretched, making it less rigid, and so reducing its transmission speed. One idea was that there's always the same amount of it, and earlier on space was more compressed. Some loose calculations got me nowhere, and I left it again.

Then in 2022 I started looking into some astronomical data that could limit the possibilities about how the time rate might be changing. The idea had sat on the shelf for so long that new technology had appeared that could shed light on it. This has happened elsewhere: the theory itself wasn't testable when I first thought of it, but it is now. If you're slow enough, the technology overtakes you, but perhaps in a good way.

People had found ways to test an effect that's very closely linked to the time rate: the cosmological redshift. The idea is that as the universe expands, light waves on their journey to Earth get stretched. Nowadays we can see distant explosions quite clearly. An event like a supernova explosion, or a gamma ray burst, which is probably from some kind of 'boom' type event, should be seen in slightly slow motion. That's due to all the different wavelengths being stretched on their way here.

The mathematics is very simple: z is the subtracted difference between true and measured wavelength (expressed in a particular way). We know the true wavelength for emission from different atoms, so we can get z. A more basic number is 1 + z, the factor difference in wavelength. Waves from a particular event all get stretched by the same factor. So 1 + z, a number we can get, should be the 'stretch factor' for space - the factor by which the universe has got larger since the light was emitted.

It also tells us about the distance at which the event happened, but there's something else. If we could film an event like that, and nowadays we sort of can, we should see it slowed down in time. This is not about a real time rate change, it's an apparent one, due to all the waves being stretched. So $1 + z$ is also the factor for how an event's duration should seem to be slowed, that is, the *apparent* time rate. You then study the curve of the explosion, and see if it's a typical one, but slowed down.

This is of interest, as finding $1 + z$ timerate adjustments at different distances is a way to confirm the expansion better than ever before. We see events at different distances or redshifts: z varies. If they're all seen 'timestretched' by $1 + z$, it strengthens the idea that the universe is expanding.

But to me the data was of interest for another reason as well: any *real* time rate changes would show up too, in this kind of measurement. And it could limit the amount by which the real underlying time rate could possibly have changed since that event happened. The $1 + z$ apparent time rate changes will slow the event down - as we see it - but any real ones should speed it up (according to this idea), as the event is seen earlier in history.

So if they were getting less than $1 + z$ here and there, it could be a real time rate difference. And reading about it, looking for it, I found that they were. A slight difference had appeared in two kinds of measurement. It was only in an average across a range of events, and within the error bars, so of course it might be nothing. But it was found in data from both supernovas and gamma ray bursts - and in each, it seemed, by the same amount.

128. Varying time cosmology (VTC)

The measurements are not all certain. In some we're looking as far as we can look, right at the edge of our instruments' range - that's part of why this idea is in the speculation section. But things can be found out, nonetheless.

In both measurements, the astronomers worked with a sample of a number of events. So because $1 + z$ varies with distance, it was natural to express the average value via an index: $1 + z$ to the power of some number. That way the results across a range of values for z could be set out. So for instance, if the apparent time rate was always $1 + z$, that would be $1 + z$ to the power of one, or $(1 + z)^1$. But what they found was just below that, the average was around $(1 + z)^{0.95}$. So the events were seen slowed, but slightly less than expected on average. That might be a faster time rate at emission, which cancels a small part of the effect they were after.

The results, from supernovas (which people call SNe) and gamma ray bursts (GRBs), were loosely speaking $(1 + z)^{0.95}$ and $(1 + z)^{0.94}$. In current physics, if this time rate difference exists, there's nothing we know of that might cause it. But in RT there are possibilities. One is the idea that when space expands by a factor x, its transmission speed is reduced, and multiplied by something like $1/(x^{0.05})$. If that were true, a lot would follow.

This alteration to the transmission speed slows the speed of light, but it also slows the background rotation, which slows time. And there's a cancellation, which leaves the travel time for light *per rotation* exactly the same. So c is still 299792458 m/sec, and the only difference is if you compare the time rate of one era with that of another. But if you do, you find that earlier in the universe's history, time ran faster, light travelled faster. You get a scenario that looks like VSL theories, or inflation. For example, the time rate T in any given era, at a redshift z, might be:

$$T_{(z)} = T_0(1 + z)^{0.05}$$

(11)

Where T_0 is the present time rate. This doesn't do much. It means that when the universe was half its present size, and z = 1, the time rate was 1.03526 times the present rate, so a little faster. Even in the early universe, you find that 50,000 years after the start, the time rate is only 1.5 times faster than at present. So this doesn't explain in any direct way the rapid star formation at high redshifts, and it's the wrong curve anyway. But if you take equation 11 further back, the time rate eventually gets enormously fast, and it is so when inflation is thought to have happened. Inflation paints a picture where things happen extremely fast at the beginning - for some reason.

The two sets of measurements were different. Only the nearby SNe data was certain enough to restrict the time rate. The GRBs went out to redshifts like 8, and different teams took the data differently. The distant flashes were *far* shorter than nearby ones. This is partly the 'tip-of-the-iceberg effect' - with very distant events, our instruments see just the peak of the explosion.

One team argued that if you adjust the data, they're all the same, but others disagreed. Another team showed, with 99% certainty, that there must be an unknown difference between the near and far ones. This leaves more room - it looked like a large time difference beyond about z = 6. Perhaps only nearby is the SNe and GRB data accurate in making it a small effect. Nearby we may be seeing the bottom of a curve that was steeper a bit further back. Before that, there could have been a rapid time rate.

Whatever you've read, cosmology is full of unknowns. There are many things for which we have no mechanism at all. The horizon problem is an example:

VSL theories and inflation were both invented partly to fix it up, but neither has a mechanism. We don't know why things happened so quickly back then, but it seems they did. What I've just described is a viable *mechanism* for it. These points sum it up and support it - points 1 and 2 are similar.

1. Space is already known to behave like a material, in other contexts. It can transmit gravitational waves (in plenty of theories including GR), and they've been detected since the 1970s. In PSG the picture is a little different, but in both pictures space can deform its shape, in one way or another.

2. We know space can probably stretch: these kind of measurements show that light waves do, and in expansion redshifts, the waves and the space they travel through seem to stretch by the same factor, $1 + z$. It looks very much like that's happening, and it would fit with other things we know.

3. If space is like a material in a particular way, you get a series of dominos in rotation theory. If space, when it is stretched, has its transmission velocity reduced - which materials often do - perhaps surprisingly, in RT the result is more or less inevitably a faster time rate in the early universe.

And that's *exactly* what measurements of early eras have been looking like. Things happened just as we know they happen, but far too quickly. So this is a possible cause, and it's in an area where we don't have causes for what we observe. It might explain a list of things we can't explain at present.

129. The accelerating expansion

Another unexplained thing is the accelerating expansion. It may turn out not to exist, but at present we have a measured anomaly. Dark energy is a name we give to the unknown cause. I'll continue telling the story now, and show how these ideas might solve that puzzle as well.

So some thinking that day I mentioned led to the ideas above, some of which had existed in less certain form already. But until that point it had been hard to find out anything about the physics of a different era.

A faster early transmission speed means things went faster, but our systems, and mine, have trouble expressing that. We can remove 'per second' - that's just a unit of ours. Instead, I use one rotation as the tick of a clock. It's also the time light takes to travel one circumference of the circle in a straight line. But because time changes as well as motion, the numeric speed is always the same - you just get c - and it's hard to crack into the physics.

I wondered what a slower background rotation rate would do to matter: if it would reduce matter's energy, which is its resistance to the flow. If the flow is slower, there's probably less resistance. But it seemed impossible to know for sure, and to know if the difference would be proportional to the speed of the flow, or go by some more complicated equation.

Then I realised (and this is why I mention a wall near a carpark) that there's a way to know, for absolutely certain, within RT. You can sometimes take what you know in one area, and apply it in another.

In a gravitational field, in general relativity, there's an unexplained reduction in energy. Matter's energy is reduced slightly when it goes near a mass. This has been measured, but there's no consensus explanation. In Book II, PSG explains it via the point that the background rotation runs slower nearer the mass, faster further away from it. The reason for that is the mass emits very small waves, which slow space's transmission speed, slowing the flow itself. Slower background rotation, less resistance from matter, so matter has less energy. What was needed in that car park was to realise that at any point in the field, matter's energy is reduced (in GR) by *exactly* the same factor as the background rotation speed is reduced (in RT). The equations are the same, I just hadn't thought about it.

So two things become clear. First, if the rotation is slower, matter needs to resist it less, just to sit there doing what it does: running to stand still. And as resistance quantifies its energy, if the flow slows down, matter's energy goes down. And what had been just a possibility was now certain: *matter's energy is exactly proportional to the speed of the background rotation*.

Apply that to cosmology, and you find the universe is doing something no-one expects it to do: it's losing mass. If all the matter in the universe is losing energy because the background rotation is slowing, then it's also losing mass in proportion, as energy and mass stay c^2 apart.

In the late 1990s we all found out, to our astonishment, that the expansion is accelerating. Two teams of astronomers discovered the acceleration in data from supernovas. It was seen as a huge discovery, and it was later confirmed by better measurements. But we've been trying to explain it ever since, with no success at all.

If the universe was losing mass, there'd be less gravity pulling it together. So the braking effect gets weaker, the expansion accelerates. This is economical as an explanation. Most of the ideas in it we know about already - things like mass and energy. There's no need for contrived ideas like quintessence, dark energy and so on. And we can't say it's the vacuum energy pushing it apart, as the numbers (via both theory and experiment), are out by 10^{120}.

This could also explain the Hubble tension - a discrepancy between two ways to estimate the expansion rate. If the universe is losing mass-energy, later on it expands faster. Ethan Siegel, in looking for solutions, points out: *There's an underlying assumption behind the expanding Universe that everyone makes, that may not necessarily be true: that the energy contents of the Universe [...] remained fundamentally unchanged as the Universe has expanded. That no type of energy has annihilated, decayed away, or transformed into another type of energy over the history of the Universe. But it's possible that some sort of energy transformation occurred in the past in a significant way.*

VTC makes predictions. One of them is: the further back you look, the faster the time rate. Another is, the time rate and the total mass stay in proportion. But to get the equation I'm searching for, the need is to see how the value of H changes as the expansion accelerates, and relate the changing mass - with clues from the time rate - to it (I'm looking for a collaborator on this).

To test rotation theory, one tests PSG, not VTC. I've set out some good ways to do that. The cosmology part is a spinoff, less certain than the main theory. But it was worth writing down, simply because it might be right.

But it's more likely to be right in the basic idea, than in the details I have right now. Equation 11 shows an example pattern: it's probably not the right one. In VTC, what we find in the early universe suggests the time rate slowed to something approximated by equation 11 later on.

If the equation is different, the interpreted GRB data will allow a faster time rate midway - between the beginning and our era. One way we might get to the real curve is via the accelerating expansion. Another way is via the time rate. And one way both the time and mass changes might be traced is via the phenomenon people used to call *downsizing*, in galaxy formation.

In the early universe, we know that large galaxies formed quicker and earlier, and then smaller ones formed slower and later. Downsizing is an important aspect of the universe, however you look at it. And it 'goes way back' as they say: it was discovered in 1996. Cosmologists have been struggling to explain it ever since. It's the opposite of our hierarchical picture. But it happened, and over billions of years. The different sized galaxy all arrived in sequence, like a series of Russian dolls.

Perhaps you can see how a faster time rate early on, plus higher mass values for all matter, can cause this. The mass changes mean stronger gravity pulls matter together into galaxies early on. The time rate adds to that, speeding up galaxy formation. Then time and mass descend, creating the sequence. A result may be what the James Webb Telescope is seeing now, which recently includes a galaxy at redshift $z \approx 14.32$, the earliest yet.

Some of the ideas here are speculative. But is the main underlying idea right: a faster rotation rate and time rate in the early universe? We know time can change its rate in two separate ways, so there may be others. Local changes to the time rate have been measured. But only RT can do that over history, in a way that might explain the observations. The mathematics and the data will confirm it or not. I'm hoping we'll find the curve, and that it fits neatly in many places. Meanwhile, it's a good chunk of jigsaw, that fits well with quite a few other parts of the picture, and it may well be right.

130. MOND, and an update on the mass discrepancy

In Book II, the gravity book, I've talked about the need to explain two sets of clues that seem to contradict each other. They support both dark matter *and* modified gravity. Each does well in its own area. MOND does well in galaxies, dark matter does well at larger scales. Many believe in one or the other, but 'hybrid theories', which manage to include both ideas somehow, have been appearing recently, and getting a lot of attention.

But all our ideas have problems. In the 'planes of satellite galaxies problem', we've found that small galaxies tend to orbit large ones in a flat plane. They shouldn't, but they do in three nearby systems. The many theories that can't explain this include ΛCDM, GR, MOND. According to PSG, a flat disk of the emitted medium builds up, bringing more satellites into that plane.

Stacy McGaugh has shown that whatever's behind the mass discrepancy in galaxies, it has a close connection with the visible matter. MOND gets at the key numbers via the visible matter (baryons), with no need for dark matter. On his site, Stacy points out the contradiction: *'We have two very different requirements on the dark matter. From a cosmological perspective, we need it to be dynamically cold. Something non baryonic that does not interact with photons, or easily with baryons. But from a galactic perspective, we need something that knows intimately about what the baryons are doing [...]. So that's where we're at right now. The two requirements are both imperative - and contradictory'.*

How do we explain a connection, *plus* a non-connection? In PSG, the visible matter emits what we call 'dark matter'. It's not matter, it's very small-scale waves in space itself, emitted by all masses, and it causes gravity. That's the connection. But once emitted, it hardly interacts with matter: that's the non-connection. At larger scales excess clouds of it build up, adding extra gravity - as in dark matter. But by then it's far from obvious where it came from.

But at smaller scales there's MOND, or modified Newtonian dynamics. Moti Milgrom published it 40 years ago, and the theory has been growing slowly, from a tiny underdog to a major, large underdog.

MOND is a theory that says gravity fields go weird at the edges. The pattern changes beyond a certain point, and the outer pattern echoes the inner one. MOND has been successful for decades, removing the need for dark matter in many places, though it has absolutely no explanation. Results that support it are often played down, but the evidence is enormous. A slight adjustment to the equations, and gravity is slightly stronger than Newton's theory where accelerations get very weak, beyond an acceleration called a_0, 1.2e-10.

MOND predicts results beforehand, without any adjustments. Dark matter is adjusted to fit results later. But they've still been in a balance: in galaxies it's hard to tell which it is. Then a weird thing happened. In 2022 new data from ESA's Gaia satellite came in, and we thought we'd now settle this once and for all. There's an arena where if we find extra gravity, it can only be MOND. Wide binaries are pairs of stars - some nearby - that orbit far enough apart to have very low accelerations at the extremes of their orbits. When the new data started to come in, it looked like MOND was going to win. And in 2023 even stronger apparent evidence appeared.

It was exciting, because if the same amount of extra gravity was found there, it couldn't possibly be dark matter, and Milgrom would - in important ways relating to gravity - replace Einstein.

It's too early to be certain, but through 2023, the ball was batted backwards and forwards. An exciting situation developed: different studies gave widely different results on wide binaries. Two teams say they proved MOND right, one says it has proved it wrong. So the jury isn't just out, some of them are way out. Still in 2023, each of the three teams then published new papers, improving their results.

Although it was two against one, the team that claimed MOND is not in wide binaries still think it works in galaxies. Indranil Banik and his team didn't find MOND in binary stars, but suggested a detailed way to adapt MOND to work at larger scales. Meanwhile, others found that in clusters, although we know from collisions there's undetected material, there's also something very like MOND, but using a different number for the limit a_0. The radial acceleration relation, or RAR, seems to be a fundamental pattern. This makes MOND look more like an effect, not a universal gravity theory.

MOND does an odd thing. It refers to Newton's acceleration values, and gets to the outer pattern of the field from the inner one. Beyond the boundary, it takes the geometric mean (near halfway) between Newton's acceleration *as it would otherwise be* there, and Newton's acceleration at the boundary, a_0. It's as if it traces a change that has happened to the field beyond a_0.

In a paper I've been writing, I've shown that in MOND it's as if beyond a_0 the field gets deformed - compressed in the radial direction. The radius term can be adjusted to preserve Newton's gravity. Or instead, in terms of the MOND radius $(GM/a_0)^{1/2}$, the simplest way to take MOND is: accelerations beyond a_0 are boosted by r/r_M. So there's a very radial aspect. The idea that the field goes through a change beyond a_0 is possible in PSG, as the field exists in its own right, like a physical object: it's made of radially travelling waves.

So as the waves travel outwards, they get affected somehow at a limit, and change their dissipation pattern. It's this pattern that causes accelerations. When they reach a_0 something *speeds up their dissipation*: they get bunched up, or start interacting with each other. Self-interaction is likely, because of what happens in clusters - the change kicks in quicker, perhaps because they come equally from all directions. In both galaxies and clusters, we know the starting point for the transition is the same: a = 2e-9. That's vital: whatever happens there, in PSG faster (altered) dissipation boosts the rate of change of the medium's state. This boosts accelerations - as MOND predicts, and as we observe. This is ongoing work, but I wanted to give you an update.

So particularly because the small-scale waves travel outwards to the edge of the field, and may encounter changes on the way, the gravity field in PSG is generally far more capable of unexpected behaviour than curved space. And that includes doing what we've found gravity does: changing the rules, and going into a different pattern, but with fascinating similarities to the inner pattern, out at the edges of the field.

131. Conclusion, and a story

Well, that's all there is for now. If you have a theory of everything, even if in places it's really just the foothills of one, there's a lot to say. I hope it was of interest, thank you for bearing with me on the way.

A lot of the main areas are in the gravity book: the construction of the theory itself, and the struggle to find evidence for it. That's the mathematical home of RT, but the conceptual home is here, with the picture of the rotation, and how it leads to the universe we know.

At times this book, with the other two, is like a 'blueprint for a paradigm', it gives a conceptual basis that can be built on. But elements of it, such as the discovery that all orbits can be described via a law from what had seemed a separate area, refraction, mean that it's far from being only a blueprint. But it still is one, and if anyone wants to try taking it further - go for it, good luck, write to me if you need anything.

If the theory goes on fitting well, in some areas it reaches there'll be people better specialised to make progress along these paths. I got to a place where the main path divides into many smaller paths. It then goes off into so many places that the best I could do in some areas, with only time to take a quick look around, was to provide a blueprint for anyone wanting to look further into these ideas, and I hope my map is easy to follow.

Now on a lighter note, here's a story about my grandmother. It's only a really good story if rotation theory turns out to be right, or a good description, and it will have to go through various tests before we'll know. But it's about how she learned about relativity, and might be of interest.

When I was very young she explained relativity to me, in a way that gave me confidence when looking at it later, and the questions surrounding it. She talked about the relative motion of boats, trains and platforms, and objects moving in empty space, and about 'flat moths', which were two-dimensional creatures trying to visualise three, just as we're three-dimensional creatures trying to visualise four.

Later on, when I truly understood relativity, I decided that neither myself nor my grandmother had understood it at all. But she was a very inspiring person nonetheless, and she had an enormous excitement and enthusiasm about it, which was infectious. Thirty years later, I found out why that was.

When she was twenty two, she married my grandfather, a writer and theatre critic in Berlin. He was in his fifties at the time, so the marriage was frowned upon by some. He was one of the first to write against the Nazis, and had to leave with his family in 1933, years before the war. (His daughter was Judith Kerr, who later became a writer, and described their travels.) They went first to France and then to Britain as refugees - because of the age difference, his wife survived him by several decades.

In Berlin before the war, one of my grandfather's friends had been Einstein, who was sometimes a visitor at the house. I found a letter Einstein wrote to Alfred Kerr in a collection of letters, and there's an account of a philosophical discussion they had in 1927, they're in the reference section. It was Einstein who explained relativity to my grandmother, when she was a young woman in her twenties. I don't know if using 'flat moths' also came from him, as part of an analogy about the dimensions.

Anyway, although I didn't understand it much at the time, being about seven or eight I suppose - for what it's worth I got it explained second hand from Einstein. One thing Einstein is quoted as having said is that you do not really understand something unless you can explain it to your grandmother. Well even in those terms it doesn't count, but he explained it to mine - and thirty five years later a rather garbled version reached me. In some ways it's not much of a story, as there must be people alive today who got it explained first hand from him later on in America, and which would probably make far better sense. But anyway, there it is.

Additional: there's an odd extra bit to the story, which I've just heard about, when almost ready to send the book in for publication. My grandmother Julia Kerr was a classical musician, and wrote two operas, one performed in Berlin. The second was about time travel, called 'Der Chronoplan' (perhaps inspired by special relativity). Her husband wrote the words - they used to write songs together. It was being prepared for the stage in 1933 when they had to leave, but in 1947, according to 'Der Spiegel' a new production was being planned, with Einstein and Bernard Shaw playing themselves. Both had agreed to do it, but Alfred died in 1948. However, I've just found out that it's to be performed again, a century after it was written - in Mainz in January 2026. Julia wasn't at all well known - it's great that they're doing it. It's coming across time, and we're looking forward to going over there to see it.

Appendix A. Antimatter

An important question in physics is: 'What is antimatter, and why is there so much more matter than antimatter?'. I'll give an answer for the first half, and a speculative one for the second. In DQM, the 'collapse of the wave function' is about the choice of a positioning for the dimensions, out of many possible positions for them implied in the wave. Before the selection, the dimensions are set at many angles, and you have a range of choices.

Now in RT the background rotation is part of the dimensions, simply because a dimension is what's rotating. So putting together ideas from DQM and RT, the selection will include a choice for the orientation of the flow.

And of all the choices for that, it could be that the geometrical possibilities include setups where the flow is rotating in the opposite direction. This extra set of possibilities would involve 'inside out' geometry, because you can't get there simply by shifting the axes around in space. But they might be valid nevertheless, and the universe will decide which geometrical possibilities are allowed, not Euclid, or anyone else - so they might be included.

And where the rotation is reversed, the circular patterns on it would be what we call antimatter: the mirror particles that have certain properties reversed from normal matter. One picture of antimatter, the Feynman-Stueckelberg interpretation, describes an antimatter particle as a normal particle travelling backwards in time. This chimes well with RT: the direction of the background rotation is very closely related to the direction of time.

For the matter-antimatter imbalance, one current guess is that almost equal amounts of matter and antimatter annihilate in the big bang, leaving a small residue behind, which led to the universe as we find it now.

In DQM, the choice of a positioning for the dimensions is limited by where and how other matter has already been slotted into a 'local framework'. It is affected by choices already made, about other nearby matter. So the answer may be that early on matter gets 'slotted into place', and once one direction for the background rotation gets ahead of the other, there's a runaway chain reaction, creating a very large majority of situations involving matter of one kind, as the structures of matter rapidly build up.

This idea is interesting, but the first part is more certain than the second. The extra set of possibilities could affect the probabilities in certain situations, so it might be possible to test it out one day.

Appendix B. The Mott problem

One of the puzzles in physics that sometimes finds its way onto lists of well-known puzzles is the Mott problem. In standard physics, it has been hard to explain for a century. No other interpretation for quantum mechanics seems to have any answer for it at all, and yet it's explained via rotation theory in a very direct way.

The solution is so direct, all you have to do it cite the DQM interpretation for quantum mechanics, and it comes straight out of the picture.

The Mott problem is this. It's found that when a quantum wave 'collapses' to the particle state in a cloud chamber, what remains is seen to be a single trail in a particular direction. Why?

Superposition in quantum theory is weird, it's hard to guess what's going on. Matter's location can be undecided. A particle can be in more than one place at once. In DQM this is because matter arises from waves in the fabric of the dimensions themselves - small disturbances that travel along the axes.

Matter exists on them. Being dimensions, they're always at right angles, but at no fixed orientation. In DQM, until a local positioning for them is picked out, the axes are set at many angles at once. Matter travels along them, so the paths it can take spread out. Matter takes them all, making a wave.

In DQM, state reduction (the rapid change from waves to particles), involves the selection of one *orientation* for the axis the particle is on, out of many. Beforehand, these possibilities are all contained in the wave - which includes a lot of paths pointing in different directions.

One of them gets picked out - that's the interpretation. It says that what is picked out takes the form of a *direction*, and a single particle. This means a single trail left after state reduction is exactly what's to be expected.

So DQM gets glimpsed in action. No other interpretation for QM says that an *orientation* is what gets picked out at the point of state reduction.

But the Mott problem boils down to the question: at the moment of state reduction (ie. wave function collapse) in a cloud chamber, why does a single trail in a particular direction remain? One answer now is simply 'DQM', but it used to be 'we don't know'. It's explained by citing the DQM interpretation, which comes out of rotation theory.

Appendix C. The principle of least action

Far and away the most important general clue is the principle of least action. It's a simple way to summarise what all matter does when it does anything. This is totally unexplained, but it turns out to be underneath all basic physics. It's an alternative way to describe what matter always does - you don't need forces, you don't need laws of motion. Instead it can all be described by light and matter doing things with the minimum possible use of energy - all the time. And it has been strongly argued, including by Richard Feynman, that whatever's going on, it's more fundamental that our laws of physics, such as for forces, laws of motion, and so on.

It even explains why matter moving in a straight line from A to B travels at a constant speed, or the path a kicked football takes. You find, amazingly, that it leads to *anything in basic physics*. So why is matter so economical? We like economy because our nature needs it, but matter isn't like that - as far as we know. There's no reason we can see, why of all the paths matter could take, it always takes the most energy-saving one, every time.

So to say this is important is an understatement. I've talked about layers of explanation (it often goes with emergence), in Chapter 39. Here it seems the principle of least action is about a deeper layer than our physics, because it can explain all of it. For instance, all three of Newton's laws of motion can be derived from it. So to explain wtf this deep layer is, there's a need to find an *even deeper* layer, perhaps with some kind of picture. If we could explain the POLA, from there we'd explain a huge number of other things.

Its history was a series of steps. First a principle of least distance (from Hero of Alexandria) then a principle of least time (from Fermat), then a principle of least action, with various versions, and contributions from Maupertuis, Euler, Lagrange, Hamilton. Could the final version be a principle of least *something else*? In rotation theory it's a principle of least resistance to the background rotation. Energy has been defined as activity, quantified by resistance to the BR. If matter is a repeating disturbance - standing waves - in a very powerful circular flow, it's clear that if matter does anything (such as moving through the flow), it's going to have to avoid resisting the flow as much as it possibly can while doing it - the flow itself will make sure that happens.

So here we have an explanation for something that underpins all our physics. It shows what a good conceptual picture can do. I've seen discussion on this at length - no-one says 'we may be missing something'. We're not geared, or taught, to think in that way. But a lateral, unexpected picture, as in rotation theory, can point down an avenue, and show a way to explain it.

Appendix D. Velcro gravity: the proof of the pudding

If you claim to have found a mathematical proof (technically a near-proof or 'smoking gun evidence') that uncovers the next paradigm, it better be good. It also better be shown - we're at the end of this series of three books, so I'll round it off with the bit of mathematics that supports the ideas in all three. I hope anyone who tries it out will find it fun, interesting or both.

You pick two points at random on any orbit through any gravity field, it then takes about ten minutes to show the present paradigm to be wrong. No-one normally proves a gravity theory, but if this picture's right, no-one has found the underlying picture before. So if it's right, it might be provable.

The picture. At a very small scale like the Planck scale, the dimensions make parallel cylinders. Light and matter are waves in their fabric, which travel in different directions through this structure. Light travels along the cylinders, matter rotates around them, both at c, the transmission speed of space for waves of all sizes. A mass like a planet, at that scale, is just a lot of rotating vibration in the dimensions themselves. So it emits weaker secondary waves that dissipate radially, making a graded medium surrounding the mass. This cloud constantly moves away and thins out, but it keeps its shape.

Matter near a mass, being like rotating light, is refracted along helical paths. As an object approaches the mass, the transmission speed of space changes faster, its helical paths get more extended, it accelerates. PSG is like 'Velcro gravity': the medium is like the wool, the helixes are like the hooks, and the mechanism only works (and is only found) if you have both components.

To test this view of gravity. Snell's law is applied to matter, assuming it is like light at a very small scale. (Equations here are numbered separately.) In well known, standard refraction, if a light beam travels through a graded or layered medium, Snell's law applies not only at a boundary, but between any two points on the light path, at any distance. For a particular beam:

$$\frac{\sin \theta_1}{S_1} = \frac{\sin \theta_2}{S_2} = \frac{\sin \theta_3}{S_3} \quad , \tag{1}$$

and so on. The S terms are local speeds for light, the θ terms are angles of the beam to the normal. So any two points on the path will behave as if they are adjacent at a boundary, and can be related via Snell's law:

$$(\sin \theta_1) / (\sin \theta_2) = S_1 / S_2 \quad . \tag{2}$$

So if numbers for two speeds and two angles are put into that equation for any pair of points on the light's path, the sides of equation 2 will agree every time, and *their agreement will show that refraction is at work.*

So far, about Snell's law, this is just standard physics. But a slight adjustment to equation 2 can be made, and it then works for orbiting matter. The two sides of the equation will agree, for two random points on any orbit - again, *their agreement shows refraction is at work.* In Planck scale gravity theory (PSG), the equivalent of a particular light beam is the helical path of a bit of matter within an object travelling through a gravity field. The helical path, if 'unrolled' to two dimensions, goes by the basic law of refraction - exactly as the light beam does. Rolled or unrolled, it's the same. So to test the theory, one simply substitutes for the four terms in equation 2.

The two S terms, local speeds for light, are replaced via expression 3, which in both GR and PSG gives the gravitational redshift. In PSG it also gives the local transmission speed of space (the local speed of light), in terms of c:

$$\sqrt{1 - (2GM/rc^2)} \ . \tag{3}$$

The two θ terms, angles to the normal, are replaced (as explained below), with the angle to the normal arccos (v/c). So substituting for some terms:

$$\frac{\sin \left(\arccos \left[v_1/c \right] \right)}{\sin \left(\arccos \left[v_2/c \right] \right)} = \frac{\sqrt{1 - (2GM/r_1 c^2)}}{\sqrt{1 - (2GM/r_2 c^2)}} \tag{4}$$

PSG is tested by putting numbers into equation 4. The sides should always agree if refraction is at work. The standard orbit speed (vis viva) equation is known to be very accurate, and its use helps in the test for that reason.

Two points are chosen on the same real orbit. It can be any orbit: elliptical, hyperbolic, parabolic - so open or closed. It can be a radial orbit, such as the path of a falling object, released from any height in a spherical gravity field. M is the central mass, G is the gravity constant. v_1 and v_2 are speeds at radii r_1 and r_2 on the trajectory, and are reached via the standard orbit speed (vis viva) equation. The two sides always agree, to $\sim$ 16 decimal places.

The Earth's field can be used, the Sun's, or any other. The quickest numbers to find online are perihelion and aphelion of planet orbits: the distances also let one calculate the speeds. The test covers a huge range of situations, and means gravity is more or less inevitably caused by refraction.

Equation 4 contains only three elements: one from gravity, the gravitational redshift, two from refraction - Snell's law and the angle to the normal. But it led to an orbit speed equation that mimics GR to 8 decimal places, and a set of other gravity equations that do the same. It also *explains* the orbits.

The only thing still to clarify is the angle to the normal, arccos (v/c). In the absence of gravity, matter's helical path involves two speed components at right angles of an overall speed c: the 3-space velocity v, and the rotational component r. In basic geometry, one loop of a regular helix, unrolled to flat, becomes a straight line - the hypotenuse of a right angle triangle. In PSG its sides are proportional to the speeds on the helical path: v, r and c. This leads via simple geometry to an angle between a straight line (the unrolled helical path) and the direction of motion: arccos (v/c). It's one of two angles related to the helix that sum to 90°, the other being arcsine (v/c).

In a gravity field, the medium thins out radially, so the normal is a radial line. When matter travels radially, this angle to the direction of motion is also the angle to the normal. Both are arccos (v/c). Equation 4 works for all orbits or paths, but it is only a near-proof where matter is on a radial path - such as, say, an apple falling from a tree. Any two points on the apple's path can be chosen, and shown to be related via the law of refraction. In the radial case, there's no need even to bring in the dimensions - all one has to assume is a graded medium surrounding the mass, and matter on helical (small-scale) paths. For why this also works on non-radial orbits, the paper '*A Planck scale theory of time*' (Kerr, 2024), mentions the local small-scale angles that arise, as matter is part of the dimensions, which can vary their orientation.

Testing by experiment

The above near-proof is strong evidence for those with an open mind, to put it mildly. But if it's right, there'll be resistance to it, and perhaps some online misinformation - GR is a major industry nowadays. (Incidentally, a pdf online summarises the mathematics of the whole theory, see the ref. section.) Still, as always, the final arbiter is experiment. At least two experiments can test PSG. They're in Book II, *Gravity: what we still don't know*, the simplest is also in the 2023 paper, *Testing a Planck scale mechanism*.

That simple experiment involves a high flying plane. You measure the speed of objects falling in a vacuum tube at different heights in the Earth's gravity field. For each measurement, calculate the Earth's mass twice from it: once using PSG, once using standard theory. Only one of the two theories will give consistent values for the Earth's mass at all heights. The other will have a pattern in it, which I've described mathematically. It's extremely subtle, but that tiny difference between GR and PSG exists.

When I started work on the theory this couldn't be measured, but that was 28 years ago, now it can. If you're slow enough, you'll find you've travelled into the future, where hopefully some people will come along and test your theory. We'll see what happens…. thanks for reading.

Index

Chapter index

References

References are listed in the format:
Chapter number [page number].

Intro [6] **Smolin**, L., *The Trouble with Physics*, Houghton Mifflin Harcourt (2006) p. 256. ISBN-10: 0618551050

Intro [7] **2009**. Hořava, Petr, *Quantum gravity at a Lifshitz point*, Phys. Rev. D 79 (8): 084008

Intro [7] (also 25 [54]) **2011**. Amelino-Camelia, G., Freidel, L., Kowalski-Glikman, J., and Smolin, L., *The principle of relative locality,* [arXiv:1101.0931]

2 [12] **Bousso**, Raphael https://www.edge.org/response-detail/11707

2 [12] **Hossenfelder**, Sabine
https://www.youtube.com/watch?v=PdL8CudJTcs *[5 mins 19 secs in]*

4 [15], (also 14 [32]) **Rietdijk-Putnam argument**, Rietdijk, C.W. (1966) *A Rigorous Proof of Determinism Derived from the Special Theory of Relativity*, Philosophy of Science, 33 (1966) pp. 341–344

Putnam, H. (1967). *Time and Physical Geometry*, Journal of Philosophy, 64, (1967) pp. 240–247

7 [20] **Muons**, Bailey et al., *"Measurements of relativistic time dilation for positive and negative muons in a circular orbit,"* Nature 268 (1977) pg 301.

8 [21] **Feynman**,1963 (4.1 - 4.2)

8 [22] **Energy article**, William Orem, *Definition: energy*, 2010
A blog article, this shows how there is no root level understanding of energy
http://fqxi.org/community/forum/topic/639

11 [26] **Discussion,** *Does light possess mass?* Quora
https://www.quora.com/Does-light-possess-mass

11 [27] **Light clock:** Gilbert Lewis, Richard Tolman, *The Principle of Relativity, and Non-Newtonian Mechanics*, Proceedings of the American Academy of Arts and Sciences (1909), 44: 709–72. Also *The Feynman Lectures on Physics*, Vol 1 (1963), also https://arxiv.org/abs/physics/0505134

11 [27] **Confined light's response to gravity,** Kerr J., *Gravity: what we still don't know*, Gordon Publishing, Bedfordshire UK (2023) ISBN 978-0-9564222-3-1

Ch 33, *The Galileo experiment, but with light and matter,* 34, *A box of mirrors*

Kerr, J. M., Phys. Ess. 36, 1, pp. 37-50 (2023)

11 [27] **T'Hooft,** M.B. van der Mark and G.W. 't Hooft, *Light is Heavy,* 2015 https://arxiv.org/abs/1508.06478

16 [36] **'Equal footing, yes; same nature, no'**, Taylor, E. F., Wheeler, J. A., Spacetime Physics: *Introduction to Special Relativity*, 2nd edn. (Freeman, New York 1992), p. 18

20 [46] **Deutsch**, David in *Physics crunch: Desperately seeking everything,* Michael Brooks, New Scientist, 8 March 2013

21 [47] **Einstein.** Einstein, A. Preface to: *Cinquant'anni di Relatività* (Fifty years of Relativity). Giuntine-Sansoni, Florence, 1955, p. XX.

21 [47] *Albert Einstein: Philosopher-Scientist*, from *The Library of Living Philosophers* Series, Cambridge University Press (1949).

21 [47] **QM complete**. Roger Colbeck; Renato Renner (2011). *No extension of quantum theory can have improved predictive power.*
Nature Communications **2** (8). arXiv:1005.5173.

24 [53] **Boorstin.** The saying appears in different forms in *The Discoverers* (1983), Random House Inc. New York, ISBN 0-394-72625-1, also *Cleopatra's Nose* (1995), and a book of quotations with that quote as the title (2020).

29 [61] **Essay paper**: *Questioning the foundations: which of our basic physical assumptions are wrong?* Kerr, J M, FQXi essay 2012 *'A short look through the clues about time'*, https://fqxi.org/community/forum/topic/1359

32 [65] **NIST**. Chou, C. W. et al, *Optical clocks and Relativity*, Science 24 Sept 2010: Vol. 329 no. 5999 pp. 1630-1633

33 [66] **Ellis**. George F. R. Ellis, Rituparno Goswami, *Space time and the passage of time.* arXiv:1208.2611v4 [gr-qc]

42 [81-2] **Turok,** Neil, welcome speech, Perimeter Inst. Recorded Seminar Archive, http://pirsa.org/displayFlash.php?id=13080001

42 [82] **Turok**, Neil, Maclean's, *Perimeter Institute and the crisis in modern physics.* Neil Turok talks to Paul Wells about the ever-increasing complexity of theoretical physics. September 5, 2013. http://www.macleans.ca/politics/

ottawa/perimeter-institute-and-the-crisis-in-modern-physics

42 [8] **Davies**, *About time: Einstein's unfinished revolution*, Paul Davies. Orion Productions 1995, Simon and Schuster, 1996 ISBN 0-671-79964-9

46 [88] **Definitions,** http://www.exactlywhatistime.com/definition-of-time

58 [111], 117 [208] **Irreversibility experiment,** T. B. Batalhao, A. M. Souza, R. S. Sarthour, I. S. Oliveira, M. Paternostro, E. Lutz, R. M. Serra, *Irreversibility and the arrow of time in a quenched quantum system,* arXiv:1502.06704v2

58 [111] **Irreversibility experiment article,** http://phys.org/news/2015-12-physicists-thermodynamic-irreversibility-quantum.html

61 [117] **Greene**, Brian, *The Fabric of the Cosmos: Space, Time and the Texture of Reality,* 2004, Knopf Doubleday Publishing Group, ISBN-13: 9780375727207

63 [122] **Gravitational waves travel at lightspeed** ApJL, 848:L12, 2017 https://arxiv.org/abs/1710.05833

66 [124-5] **Kaluza**, Theodor (1921). *Zum Unitätsproblem in der Physik*, Sitzungsber. Preuss. Akad. Wiss. Berlin. (Math. Phys.): 966–972.

66 [125] **Klein**, Oskar (1926). *Quantentheorie und fünfdimensionale Relativitätstheorie,* Zeitschrift für Physik A **37** (12): 895–906.

75 [136] **Einstein**, Albert (1924) 'Über den Äther', *Verhandlungen der Schweizerischen Naturforschenden Gesellschaft* 105:2, 85-93. Also published in English: S.W. Saunders, translator. *The Philosophy of Vacuum*, edited by Simon Saunders and Harvey R. Brown, pp. 13-20; Clarendon Press, Oxford, 1991. ISBN 0-19-824449-5. By permission of Oxford University Press.

76 [140] **Penrose**, R, (1991). *"The mass of the classical vacuum"*. In S. Saunders, H.R. Brown, *The Philosophy of Vacuum* (a collection of essays) Clarendon Press, Oxford, 1991. ISBN 0-19-824449-5. Also by permission.

96 [175]. **RQM.** Rovelli, C., *"Relational Quantum Mechanics";* International Journal of Theoretical Physics **35**; 1996: 1637-1678; arXiv:quant-ph/9609002

100 [180] **Kochen**, Simon, Symposium of the Foundations of Modern Physics: 50 Years of the Einstein-Podolsky-Rosen Gedanken experiment (World Scientific Publishing Co., Singapore, 1985), pp. 151–69.100

[181] **Hilbert,** David. P. Frank, *Einstein - His Life and Times*, p. 206.

108 [195] **Penrose**, R. (1989). *The Emperor's New Mind*, ISBN: 0-19-851973-7, Oxford University Press

109 [198] **Pions.** T. Alvaeger et al. *Test of the second postulate of special relativity in the GeV region*, Physics Letters A, vol. 12, no. 3, 260-262, 1964

110 [200] Kerr, J. M., *The Unsolved Puzzle: Interactions, not measurements,* Gordon Publishing, Bedfordshire, UK (2019) ISBN 978-0-9564222-6-2

110 [200] Kerr, J. M., *Gravity: what we still don't know*, Gordon Publishing, Bedfordshire UK (2023) ISBN 978-0-9564222-3-1

110 [200] Kerr, J. M., Phys. Ess. 33, 1, pp. 1-9 (2019)

110 [200] Kerr, J. M., Phys. Ess. 36, 1, pp. 37-50 (2023)

113 [203-4] **Wheeler quote** (rederiving quantum theory). Transcript of a 1980s BBC radio interview, *The Ghost in the Atom: A Discussion of the Mysteries of Quantum Physics*. J. R. Brown [edited], P. C. W. Davies, Cambridge University Press, 1986. ISBN: 9780521307901

126 [222], 129 [229-30], **Downsizing** F Fontanot et al., *The many manifestations of downsizing: hierarchical galaxy formation models confront observations,* MNRAS, vol 397, 4, August 2009, pp 1776–1790

126 [222], 129 [229] **Rapid star formation** R. Decarli, et al., *Rapidly star-forming galaxies adjacent to quasars at redshifts exceeding 6,* Nature vol 545, pp 457–461 (2017) **Article:** https://phys.org/news/2017-05-fast-growing-galaxies-early-universe.html

128 [225-6] **Supernovas** Blondin, S., et al., *Time Dilation in Type Ia Supernova Spectra at High Redshift*, Astrophys.J. 682 (2008) 724-736 https://arxiv.org/abs/0804.3595

128 [225-6] **Gamma ray bursts** Fu-Wen Zhang et al., *Cosmological time dilation in durations of swift long gamma-ray bursts,* Astrophys. J. Letters, 778:L11 (5pp), Nov. 2013

128 [226] **Gamma ray bursts** O. M. Littlejohns, N. R. Tanvir, R. Willingale, P. A. Evans, P. T. O'Brien, A. J. Levan, *Are gamma-ray bursts the same at high redshift and low redshift?* MNRAS, vol 436, Issue 4, 21, 2013, pp. 3640–3655

129 [229] Stacy McGaugh's Triton Station blog, *Old galaxies in the early universe*

130 [230] **MOND** Milgrom, M, ApJ **270**, 371 (1983)

130 [231] **Radial acceleration relation (RAR) curve extended,** Brouwer et al., Astronomy and Astrophysics **650**, A113 (2021), arXiv:2106.11677v1
https://tritonstation.com/2021/06/28/the-rar-extended-by-weak-lensing

130 [231] **Wide binaries** X. Hernandez, V. Verteletskyi, L. Nasser, A. Aguayo-Ortiz, arXiv:2309.10995v2 (2023)

130 [231] **Wide binaries** Chae, Kyu-Hyun, arXiv:2309.10404v4 (2023)

130 [231] **Wide binaries** Indranil Banik, Charalambos Pittordis, Will Sutherland, Benoit Famaey, Rodrigo Ibata, Steffen Mieske, Hongsheng Zhao, *Monthly Notices of the Royal Astronomical Society*, Volume 527, Issue 3, pp. 4573–4615 (2023)

130 [231] Yong Tian et al., *A distinct radial acceleration relation across brightest cluster galaxies and galaxy clusters*, arXiv:2402.12016v1 (2024)

130 [232] Thanks to Nigel Howe for the use of the photo

131 [234] **Einstein to Alfred Kerr letter,** 1921
https://einsteinpapers.press.princeton.edu/vol12-trans/90

131 [234] **Discussion,** from the diary of Count Kessler, describing events at a dinner at publisher Samuel Fischer's home in Berlin, 14th Feb 1927. Kessler, p36. Discussion (in which Alfred Kerr is jokingly derisive and Einstein behaves with 'great dignity'), p39-40. http://press.princeton.edu/chapters/s6681.pdf

131 [234] **Der Chronoplan** *'George Bernard Shaw and Albert Einstein will appear on stage as characters in a German opera.'* Der Spiegel, Jan 1947. https://www.spiegel.de/kultur/drei-beruehmte-und-eine-oper-a-8cbe2b0c-0002-0001-0000-000038936594

Appendix B [236] **The Mott problem** Nevill Mott, *The Wave Mechanics of α-Ray Tracks*, Proceedings of the Royal Society (1929) A126, pp. 79-84

Appendix C [237] **Principle of least action** N. S. Manton, Cambridge, Sections 4-7. https://www.damtp.cam.ac.uk/user/nsm10/PrincLeaAc.pdf

Appendix D [238] **Summary of the main mathematics** of Planck scale gravity theory (PSG), the area of rotation theory where most new mathematics is to be found. Elsewhere the mathematics of RT is mainly rederivations. https://gwwsdk1.wixsite.com/link/summary-pdf

www.ingramcontent.com/pod-product-compliance
Lightning Source LLC
Chambersburg PA
CBHW071633200326
41519CB00012BA/2285